Next-Generation Wargaming for the U.S. Marine Corps

Recommended Courses of Action

YUNA HUH WONG, SEBASTIAN JOON BAE, ELIZABETH M. BARTELS, BENJAMIN SMITH

NATIONAL DEFENSE RESEARCH INSTITUTE

Prepared for the United States Marine Corps
Approved for public release; distribution unlimited

For more information on this publication, visit www.rand.org/t/RR2227

Library of Congress Cataloging-in-Publication Data is available for this publication.
ISBN: 9781-9774-0311-7

Published by the RAND Corporation, Santa Monica, Calif.
© Copyright 2019 RAND Corporation
RAND® is a registered trademark.

Support RAND
Make a tax-deductible charitable contribution at
www.rand.org/giving/contribute

www.rand.org

Preface

This report gathers information on the tools, approaches, best practices, and other lessons learned from a wide variety of organizations involved in defense or national security wargaming, in order to make recommendations to the U.S. Marine Corps as it seeks to expand its wargaming capability. This research was sponsored by the Marine Corps Operations Analysis Directorate and was conducted within the Navy and Marine Forces Center of the RAND National Defense Research Institute, a federally funded research and development center sponsored by the Office of the Secretary of Defense, the Joint Staff, the Unified Combatant Commands, the U.S. Navy, the U.S. Marine Corps, the defense agencies, and the defense intelligence community.

For more information on the Navy and Marine Forces Center, see https://www.rand.org/nsrd/ndri/centers/navy-and-marine-forces.html or contact the director (contact information is provided on the webpage).

Contents

Figures

Tables

Summary

The purpose of this report is to provide the U.S. Marine Corps with recommendations and courses of action (COAs) as it invests in its wargaming capability. The Marine Corps wishes to emphasize wargaming to support concept development, wargaming to support capability development and analysis, and strategic leader engagement and strategic discussion. Because of the possibility that Marine Corps wargaming might be called upon to broaden its scope in the future, we also reviewed three other categories of wargaming: science and technology wargaming, wargaming to support operational decisions and plans, and wargaming to support training and education. The Marine Corps also has a future vision for wargaming in which it wishes to remove some of the current limitations in gaming, as outlined in its next-generation wargaming (NGW) concept.

This report builds on a previous task for the Marine Corps Wargaming Division (WGD) in which RAND Corporation conducted interviews at a number of wargaming centers outside the Marine Corps in order to gather information on the tools and approaches they used for wargaming. We also collected information on wargaming processes, facilities, skill sets, and other recommendations, and provided summaries of the organizations we contacted for this report.

While our previous effort produced a catalog of tools and approaches, for this report we wanted to provide the Marine Corps with a better way to evaluate the utility of different wargaming tools in relation to building its NGW concept. With this in mind we identified tasks by wargaming type in order to provide additional information on when in the wargaming process certain tools might be useful. Table S.1 lists the tasks that were common to all the wargaming categories we examined. Our intent is to make the catalog of tools and approaches more useful by allowing readers to see that there is a task, such as managing information, and then allow them to see what tools different wargaming centers may use for that purpose. In the report we also discuss additional tasks and the tasks that are particularly emphasized for different categories of wargames. This is to allow readers to see where there may be specialized tasks for categories such as concept development or senior leader engagement and discussion, and note if any tools and approaches may fit there.

Table S.1
Core Wargaming Tasks

Core Wargaming Tasks	Description
Understanding sponsor requirements	Understanding the sponsor's objectives for the wargame and analysis.
Managing information	Acquiring and coordinating research among the wargame team.
Understanding the problem	Framing the problem to inform game design.
Developing and managing the event process	Coordinating logistical concerns involved with the game such as venue, participants, and classification.
Scenario development	Crafting a scenario designed to inform and draw out the necessary decisions for participants within a game.
Game design	Creating and developing a system of decisionmaking and consequences within the game.
Game development	Developing a game through scenario validation, rule development, playtesting, and refinement.
Providing facilitation	Facilitating the game to keep players engaged and game events or discussion progressing.
Data capture	Capturing data and insights from the wargame, either through note takers or other tools.
Wargame analysis	Analyzing the wargame's insights, results, and data.

Although we gathered information about the different tools and approaches, both manual and software-based, that the different wargaming centers used, our sponsor wanted to emphasize software tools in our analysis. This is not to say that software-based tools are inherently superior to manual techniques and tools, or vice versa. There are advantages and disadvantages to both, as outlined in further detail in Chapter 5. However, in response to sponsor needs, our analysis emphasized software-based tools and categorized them into four primary "ecosystem" categories as a way to organize the software tools that centers were using for wargame support. The different software tool ecosystems are listed in Table S.2, and we also refer to these in our COAs.

While we leave it to the Marine Corps to experiment with and adopt specific, individual software-based tools, we believe that categorizing them into these ecosystems may make some of their advantages and disadvantages more explicit. Furthermore, unlike manual techniques or tools, software-based tools pose unique challenges in terms of information infrastructure, cost, maintenance, and utility. Subsequently, this requires long-term vision and coherency when building any software-based capability on a large scale.

The Marine Corps also asked us for low-, medium-, and high-resourced COAs. Our estimate of what constitute low-, medium-, and high-resourced COAs is approximate, as a detailed cost estimate was beyond the scope of this study. Rather than

Table S.2
Software Tool Ecosystems

Primary Ecosystem	Advantages	Disadvantages
Commercial off-the-shelf (COTS)	• Larger user community • Customer support available	• Not all are allowed on government networks • Less customization • Less-developed data or scenarios for the military
Program of record[a]	• Common reference point across multiple communities • Often existing data • Existing users	• Reduced flexibility • Less used in wargaming • Some require large footprint (facilities/people)
Government off-the-shelf (GOTS)	• Already developed • Free to government users • Existing users	• Requires knowledge of tool • Sometimes limited user community
Custom-built	• High degree of customization possible	• Requires time and money to develop • Redundancy with existing tools

[a] A program of record is a program that is funded (approved) across the Future Years Defense Program through the Program Objective Memorandum.

presenting the COAs and recommending one, however, we approached them so that they would build upon one another. The reason we did this is that we realize there are relatively low-cost actions that the Marine Corps would be able to take in any case, regardless of the amount of resources available. We also realize that building up a future wargaming capability is not an overnight process, and that certain steps build on others in a progressive fashion.

Within our COA recommendations, we allotted recommendations to low-, medium-, and high-resourced COAs through an estimation of cost, as defined by time of implementation, complexity, and risk as in scale:

- Recommendations in the **low-resourced COA** are actions that can be implemented with minimal resources and relatively quickly. These focus on improving processes and the wargaming skills of the staff already engaged.
- The **medium-resourced COA** builds on the low-resourced COA and makes recommendations that require more resources, time, and research and involve a greater degree of uncertainty. These focus on acquiring additional skill sets, personnel, and equipment.
- The **high-resourced COA** builds on both the previous COAs and requires major reorganization or major changes in policy or actions of inherent high complexity and financial commitment. Assuming a high level of resources, these recommendations focus on the construction of new and specialized wargaming facilities.

Again, the COAs are meant to build on each other rather than representing choices between discrete options.

Recommendations for the Low-Resourced Course of Action

1. Continue Marine Corps–level integration of wargaming with capability development and programmatic decisionmaking.
2. Focus on rapid- or quick-turn games that can explore issues in a responsive way by leveraging small groups and flexible game designs. This process can be supported by contracting support by labor time rather than a set number of games to provide operational flexibility to the center.
3. Bring in additional low-cost resources to support innovative gaming. Based on practices at other wargaming centers, these might include spreadsheet tools, materials to create board games and to modify commercial board games, and a collection or library of commercial games and wargaming publications (magazines, journals, research, and books).
4. Begin a GOTS ecosystem through the adoption of one of the many GOTS wargame support tools available. We recommend experimenting with different tools and adapting and further developing them for Marine Corps purposes.
5. Provide formal wargaming education to staff on different wargame design methodology, styles, and adjudication methods.
6. Share information and collaborate with other wargaming organizations, through formal staff exchanges, informal inclusion in other organizations' games, and inviting other organizations to Marine Corps games.
7. Establish and expand a network of experts to draw into wargames, including faculty at the Marine Corps University and subject matter experts available at universities, research centers, think tanks, and other organizations in the greater Washington, D.C., metropolitan area.

Recommendations for the Medium-Resourced Course of Action

1. Bring in additional staff with skill sets in software development or programming; information technology and network support; operations research, social science, or other analytical backgrounds; and wargame design experience with other defense organizations or commercial firms.
2. Add staff and equipment for multiple networks, including a stand-alone network that is only for WGD use and not connected to other networks. The other networks may include the Non-Classified Internet Protocol Router Network, the Secret Internet Protocol Router Network, and the Joint Worldwide Intelligence Communications System.

3. Provide staff time to further develop existing tools and to research and prototype with new gaming constructs.
4. Provide Top Secret Sensitive Compartmentalized Information clearances for staff in order to support events at this higher level of classification.
5. Add COTS tools for knowledge management, data capture, analysis, video creation, and graphic design.
6. Buy a large-format printer, a tool that many wargaming centers find useful in creating maps and visuals for their wargames.

Recommendations for the High-Resourced Course of Action

1. Consider the highest level of classification that the Marine Corps desires to handle in its wargames as it builds new wargaming spaces. Have at least part of any new wargaming facility cleared at this level.
2. Build configurable spaces with movable and reconfigurable seating, tables, and walls. This allows maximum flexibility.
3. Build configurable networks and configurable audiovisual equipment that can support configurable spaces.
4. Create workspaces for wargame teams, apart from spaces for wargame participants.
5. Furnish a reference library. Allocate a physical space for the library to store commercial board games, games on personal computers, modified games, books, game construction materials, and other reference materials for wargames.

We recommend that the Marine Corps implement as many COAs as possible, depending on the resources available, and beginning with the low-resourced COA. If the Marine Corps has resources enough for constructing wargaming facilities, we recommend that it implement all three COAs in sequential order.

The Marine Corps currently faces an environment conducive not only to adopting best practices, tools, and approaches from others but also to adapting and developing them further to suit Marine Corps needs. A number of factors currently make this environment favorable: the current Department of Defense emphasis on wargaming, continued improvements in wargaming education, opportunities to learn from other defense wargamers, the number of wargaming support tools currently available, and ongoing developments in commercial gaming. However, time for Marine Corps wargamers to further develop skills and to experiment and adapt existing approaches will be key.

Acknowledgments

We would like to thank the sponsor and several organizations at the Marine Corps who assisted us and provided valuable direction during the course of this study. The Operations Analysis Directorate (OAD), Marine Corps Systems Command, and others were especially helpful during the course of our work. We are particularly indebted to William Inserra, John Lawson, and Bret Telford at OAD; and to William Lademan at the Wargaming Division.

We are also well aware that without the valuable information provided by wargamers across the organizations we contacted, we would not have had much of value to report back to the Marine Corps. Many thanks are due the dozens of individuals we spoke with in the wargaming community for their generosity with their time and their knowledge. We greatly appreciate their willingness to speak with us during our center and tool write-ups—at times a very time-consuming process. This report is possible because of them.

We would also like to thank the RAND Corporation management team that oversaw this study and provided feedback: Mike Decker, Paul DeLuca, Seth Jones, and Mike McNerney. We would also like to thank our reviewers for this study: Peter Perla at the Center for Naval Analyses, and John Yurchak at RAND.

Abbreviations

ACH	analysis of competing hypotheses
ACSC	Air Command and Staff College
AFMC	Air Force Materiel Command
AFRL	Air Force Research Laboratory
AR	augmented reality
ARCIC	Army Capabilities Integration Center
AV	audiovisual
AWARS	Advanced Warfighting Simulation
CAA	Center for Army Analysis
CAEn	Close Action Environment
CAPE	Cost Assessment and Program Evaluation
CASL	Center for Applied Strategic Learning
CCMD	Combatant Command
CGSC	Command and General Staff College
CNA	Center for Naval Analyses
COA	course of action
COIN	counterinsurgency
CONOPS	concepts of operation
COP	common operational picture
CoP	community of practice
CORA	Center for Operational Research and Analysis

COTS	commercial off-the-shelf
CSL	Center for Strategic Leadership
DME	dynamic morphological exploration
DoD	Department of Defense
DSE	Directorate of Simulation Education
DST	Defence Science and Technology (Australia)
Dstl	Defence Science and Technology Laboratory (United Kingdom)
DSW	Department of Strategic Wargaming
DWAG	Defense Wargaming Alignment Group
EW	Electronic warfare
FAST	Future Analytical Science and Technology
GCM	Green Country Model
GMA	general morphological analysis
GOTS	government off-the-shelf
GUI	graphical user interface
ISIL	Islamic State of Iraq and the Levant
ISR	Intelligence, Surveillance, and Reconnaissance
IT	information technology
JHU-APL	Johns Hopkins University Applied Physics Laboratory
JLASS	Joint Land, Aerospace and Sea Strategic Exercise
JPME	joint professional military education
JSAF	Joint Semi-Automated Forces
jSWAT	Joint Seminar Wargaming Adjudication Tool
jSWAT2	Joint Seminar Wargaming Adjudication Tool 2
JWAM	Joint Wargame Analysis Model
M&S	modeling and simulation
M&SCO	Modeling and Simulation Coordination Office

MANA	Map Aware Non-Uniform Automata
MAPEX	map exercise
MARCORSYSCOM	Marine Corps Systems Command
MCCDC/CD&I	Marine Corps Combat Development Command / Combat Development and Integration
MCU	Marine Corps University
MOD	Ministry of Defence
MORS	Military Operations Research Society
NATO	North Atlantic Treaty Organization
NDU	National Defense University
NGW	next-generation wargaming
NIPRNet	Non-Classified Internet Protocol Router Network
NPS	Naval Postgraduate School
NUWC	Naval Undersea Warfare Center
NWC	Naval War College
OAD	Operations Analysis Directorate
ONA	Office of Net Assessment
ONI	Office of Naval Intelligence
OPLAN	Operation Plan
OR	operations research
OSD	Office of the Secretary of Defense
PC	personal computer
PME	professional military education
PMESII-PT	Political, Military, Economic, Social, Information, Infrastructure, Physical Environment, and Time
POR	program of record
PPBES	Planning, Programming, Budgeting, and Execution System
PSOM	Peace Support Operations Model

RCAT	Rapid Campaign Analysis Toolset
RFLEX	RAND Framework for Live Exercises
S&T	science and technology
SAGD	Studies, Analysis, and Gaming Division
SAP	Special Access Program
SATs	structured analytic techniques
SCI	Sensitive Compartmentalized Information
SimBAT	Simulation-Based Analyst Training
SIPRNet	Secret Internet Protocol Router Network
SME	subject matter expert
SOCOM	Special Operations Command
SOD	Systemic Operational Design
SOF	Special Operations Forces
SPSS	Statistical Package for Social Sciences
SSA	Support for Strategic Analysis
SSM	soft systems methodology
SSR	Synthetic Staff Ride
SWIFT	Standard Wargaming Integration and Facilitation Tools
TRAC	The Research and Analysis Center
TRAC-FLVN	TRAC Fort Leavenworth
TRAC-WSMR	TRAC White Sands Missile Range
TRADOC	Training and Doctrine Command
UFMCS	University of Foreign Military and Cultural Studies
USAWC	U.S. Army War College
VAST	Versatile Assessment Simulation Tool
VR	virtual reality
VTC	video teleconferencing
WGD	Wargaming Division

Glossary

The U.S. national security wargaming community has many terms of art that may not be immediately understandable to outside observers. Some of the most commonly used terms are listed here and used throughout this report. Although the wargaming community does not always agree on terminology, the definitions listed in Table G.1 are intended to reflect the most commonly held meaning for the terms listed. There is more than one definition in some cases.

Table G.1
Common Wargaming Terms and Definitions

Term	Definition
Adjudication	The method of determining the outcome of (often competing) player actions in a game.
Blue cell	The players in a game representing friendly forces.
BOGSAT	Shorthand for "Bunch of Guys Sitting Around a Table." Used pejoratively to indicate a lack of systematic adjudication.
Counter	A cardboard square or rectangle used to represent forces on a game map. Counters often display U.S. military operational terms and graphics.
Expert adjudication	An adjudication style that relies on human expert judgment to determine the outcome of player actions in a game. Also called free adjudication.
Green cell	1. The players in a game representing allied forces. 2. The players in a game representing civilians rather than military forces.
Hex game	A game where the physical representation of terrain and the movement of representative forces is along a hexagonal grid. Developed first by the RAND Corporation in the 1950s.
Hot wash	Facilitated, after-game discussion where game participants offer insights and provide feedback on the game.
Kriegsspiel	The original nineteenth-century wargame developed by the Reisswitz family for the Prussian military. Widely regarded as the beginning of modern wargaming.
Matrix adjudication	An adjudication style that relies on player arguments for expected success or failure of player actions.

Table G.1—Continued

Term	Description
Matrix game	A game, often multisided, that uses matrix adjudication. Developed first by the American social worker Chris Engle.
Miniatures game	A game, often based on historic events, that uses ornate miniature physical replicas of the forces involved.
No-turn game	A game without set turn-taking. Actions and adjudication are as near to real time as possible.
Program of record	A program that is funded (approved) across the Future Years Defense Program through the Program Objective Memorandum.
Red cell	1. Defense community definition: the players in a game representing adversary forces. 2. Intelligence community definition: an independent group that challenges and offers alternatives to an organization's standard processes and assumptions.
Red team	1. Defense community definition: an independent group that challenges and offers alternatives to an organization's standard processes and assumptions. 2. Intelligence community definition: the players in a game representing adversary forces.
Rigid adjudication	An adjudication style that relies on preset probabilities to determine the outcome of player actions.
Seminar game	1. A wargame format where players openly discuss their sequence of moves and countermoves in a given situation. The format is designed to elicit insights through guided discussion and discourse. Focus is on group conversation about what decisions ought be made and why, indicated primarily through player statements. Generally, there is a strong focus on player roles and scenario rather than on rules or adjudication model. 2. A facilitated group discussion, similar to what is found in a seminar classroom, where participant discussion rather than lecture is the dominant means of learning. 3. A game where actors specify their actions and the events that happen rather than involving others as adjudicators.
Tabletop exercise	A facilitated discussion around a specific topic of interest. The most common form is similar to that of a seminar classroom discussion rather than a set of adjudicated actions.
Title 10 wargame	A major service–sponsored wargame that addresses future concepts and capabilities in the context of Title 10 responsibilities to organize, train, equip, prepare, and maintain the services forces to carry out its roles and functions.[a]
Wargame or game	1. A model or simulation of warfare, not involving actual forces, in which the flow of events is affected by, and in turn affects, decisions made during the course of those events by players representing the opposing sides.[b] 2. One of a wide variety of facilitated defense community activities, including discussions, planning activities, exercises, and adjudicated simulations of warfare.
White cell	The individuals who run and adjudicate a game. Also called control.

[a] Douglas Ducharme, "Approaches to Title 10 Wargaming," Newport, R.I.: Naval War College, Wargaming Department, undated, p. 1.

[b] Peter Perla, *Peter Perla's The Art of Wargaming: A Guide for Professionals and Hobbyists*, ed. John Curry, Morrisville, N.C.: Lulu Press, 2012, p. 280.

CHAPTER ONE

Introduction

Research Objectives

The primary purpose of this study is to provide recommendations for the U.S. Marine Corps as it seeks to expand its wargaming capability by considering a new wargaming center. The RAND Corporation consulted Marine Corps stakeholders, including the Operations Analysis Directorate (OAD), the Wargaming Division (WGD), and Marine Corps Systems Command (MARCORSYSCOM) in developing low-, medium-, and high-resourced courses of action (COAs), as defined by monetary cost, time, complexity, and difficulty of execution. This project leverages a previous RAND effort that compiled a catalog of wargaming tools and approaches that other wargaming centers use in wargaming. Another study objective has been to document these previous interviews at other wargaming centers in order to discuss additional information that these centers offered about best practices, skill sets, processes, facilities, and other topics.

The focus of the new wargaming capability is to support concept and capability development and to enable senior leader and strategic discussions. Although the Marine Corps University (MCU) is an important player in wargaming to support education, recommendations on educational wargaming in the Marine Corps was considered outside the scope of this research. Detailed cost estimates for the recommended COAs were also outside the scope of the study. This report is instead intended to assist the Marine Corps as it makes decisions about developing its future wargaming capability during a time of U.S. Department of Defense–wide interest in expanding and improving wargaming. Although this report was specifically designed for the Marine Corps, we believe its analysis and catalog can add value to the wider wargaming community.

Tasks

The tasks for this study were to (1) document the best practices and recommendations already offered at other wargaming centers during previous interviews; (2) evaluate the current state of the art for wargaming; (3) develop COAs for the Marine Corps as it expands its wargaming capability; and (4) provide a final report that includes a catalog of tools, approaches, best practices, and recommendations.

Approach

In this report we provide context about the recent increased interest in wargaming within the U.S. Department of Defense (DoD) and give a current description of the defense wargaming community. We do this by discussing the high-level guidance on reinvigorating wargaming in DoD and through additional interviews with wargamers involved in improving knowledge and practice within DoD.

We also consulted our interview notes and emails from the numerous exchanges we previously had with wargaming centers as we compiled a catalog of wargaming methods and tools. We did so to document the best practices and insights about processes, skill sets, facilities, and other issues that are not necessarily captured in a tools and approaches catalog. We also consulted briefings, handbooks, guides, articles, and other publications that these organizations had produced about wargaming.

The selection of the wargaming centers that we contacted was based on consultations with the sponsor, our existing knowledge of DoD and allied wargaming organizations, and our efforts to get a sample from across these different organizations. For example, we aimed for interviews through the services, the Combatant Commands (CCMDs), the Office of the Secretary of Defense (OSD), federally funded research and development centers, and wargaming allies such as Australia, Canada, and the United Kingdom. The sponsor approved the list of organizations before our interviews began. We excluded Marine Corps wargaming organizations because the sponsor decided that it already had access to these organizations and was knowledgeable about their activities. Marine Corps stakeholders also had specific requests about which organizations to visit or conduct interviews at. A few organizations were consulted about a specific tool and did not have wargaming recommendations beyond the tool, and that information is included in Appendix B.

In consultation with our Marine Corps sponsors, we also developed descriptions of several categories of wargaming. Input from MARCORSYSCOM, OAD, the WGD, and others helped us to identify and refine these categories. By building out the characteristics of categories in use by the Marine Corps and connecting them to broader concepts in the wargaming community, we have aimed to create a set of consistent, clear terms that can describe different types of game purposes and then use these categories in the more general report.

We then built a framework that identified the tasks that each category of wargaming required, and referenced the information in the previously compiled tools and approaches catalog, along with the recommendations and best practices that other organizations passed along. We did this based on processes and tasks identified by different centers when discussing their wargaming activities, on published DoD wargaming handbooks, on several center briefings about their wargaming processes, and on the study team's knowledge of wargaming tasks from our own experience with wargaming. We also shared these frameworks with the sponsor and certain wargaming centers before drafting our recommendations and report.

Focusing on the categories of wargaming that the Marine Corps stakeholders identified as important to support, we then built COAs for Marine Corps wargaming. The WGD requested low-, medium-, and high-resourced COAs. We put together the COAs to build on one another: the low-resourced COA encompasses the actions that the Marine Corps could take with minimal additional resources, focusing mostly on process; the medium-resourced COA adds steps that could be taken were some more resources available, mostly looking at additional skill sets and equipment; and the high-resourced COA then adds the actions that could be taken if a very high level of resources were available, mostly looking at additional facilities. We based these COAs on the best practices and lessons discussed by members of the various wargaming centers that we interviewed. However, we also considered the future vision that the WGD has expressed, so as to not simply re-create the present state of the art, as the WGD also requested.

The Organization of This Report

This report is organized to present the information we gathered during the course of our research, as well as to lay out our recommendations for Marine Corps wargaming. Chapter 2 defines the scope of the study by defining wargaming, providing a brief history of the technique, and presenting the categorization of wargaming that this report examines. Chapter 3 provides an overview on the history and current context of wargaming, both within the DoD and in the broader context of gaming. It is within this context that the Marine Corps is expanding its wargaming capabilities. Chapter 4 provides the Marine Corps' vision for next-generation wargaming (NGW) and how it has been informed by generational constructs in gaming. Chapter 5 provides the methodology and overview for our interviews with wargaming center members, as well as subsequent major findings, recurring themes, and an assessment of the state of the art in wargaming. Chapter 6 outlines the framework of tasks associated with each category of wargaming, as stated in Chapter 2, and the ecosystem of software tools utilized in this report. Based on this framework and the best practices of other wargaming centers, Chapter 7 presents our recommended COAs, a three-tier process, defined by the demand for resources, as well as key takeaways and next steps.

Appendix A contains write-ups for most of the wargaming centers at which we conducted interviews, and is a result of drafts shared with each center. Appendix B contains the complete catalog of tools and approaches collected during this project. Appendix C contains complete versions of the tables of wargaming tasks found in Chapter 6.

CHAPTER TWO

Defining Wargaming

Before diving into a description of the current state of the field, and recommendations for the future of wargaming, it may first be helpful to define wargaming for the purposes of this report. This chapter offers our definition of wargaming, defines several categories of the tool, and offers a brief history of the use of games. The chapter concludes with a brief discussion about the utility of gaming. Our hope is to help situate an unfamiliar reader, and to provide context on the authors' perspective for a more experienced reader.

What Is a Wargame, and What Is It Used For?

A wargame involves human players or actors making decisions in an artificial contest environment and then living with the consequences of their actions. Games consist of actors who make decisions, an environment they seek to effect, rules that govern what decisions they can make, and adjudication models that specify how actions affect both actors and the environment. These broad contours mean that different games can look quite different, depending on the setup of each of these four elements. Actors can be defined by detailed role descriptions or by the understanding of the players. Environments can be represented by specific maps or a few paragraphs of written scenario. Rules can be implemented by rigid mechanics or simple reminders of existing authorities and permissions. Models can be extensive computer simulations or the unstated mental models of experts. The art and science of game design is selecting a combination of these features that creates a model that is most useful in achieving the purpose of the game.

The broad flexibility of this consensus definition has long caused different gamers to draw the line between what is and is not a game. For example, some designers hold that if human players are not the center of events, the activity is more properly defined as a *model* or *simulation* rather than a game. Similarly, some argue that if real forces are involved in play, the event is no longer artificial, and has become an exercise or maneuver. Still others stress that if players are not pushed to make clear decisions and live with the consequences of their actions, then what is being conducted is a seminar or workshop rather than a game. A host of distinct terms have been invented to describe different points along these spectrums, with little consensus as regards definitions. In the interest of inclusion, we have opted to be very broad in our adjudication of what is or is not a wargaming approach or tool on the grounds that it is easier for a reader to discard an approach he or she feels is inappropriate than to guess what we might have included given a different set of criteria.

Categories of Wargames

There are any number of ways to categorize the events and activities that fall under the title of wargame within the defense community. It is possible to create typologies of wargames by purpose, such as the goal of a game. Table 2.1 is an example of one such typology that categorizes games by the goal of the game versus the type of problems games address:

It is also possible to categorize games by their adjudication style: seminar (non-adjudicated), matrix, expert, and rigid.[1] Yet another way of categorizing wargames is by level of analysis, such as strategic, operational, and tactical. All are useful ways of thinking about the games.

For the purposes of this report, we take a slightly different approach to categorizing games. In consultation with the study sponsor, we use a typology that arranges types of games by their organizational purpose—largely within U.S. military context and specifically with an eye toward U.S. Marine Corps wargaming. Some gaming centers focus on one or a few of these because they are the types of games best suited to support their larger organizational mission. While this is not a typology that may be applicable for all gaming organizations, it is the most appropriate approach to organizing insights that could benefit the sponsor. This is also an approach that identifies archetypes of games that currently exist within DoD rather than an exhaustive classification of game types using strict definitions.

The six types of wargames we have identified are: (1) wargaming for concept development; (2) wargaming to support capability development and analysis; (3) science and technology (S&T) wargaming; (4) senior leader engagement and strategic discussion; (5) wargaming for operational decisions and plans; and (6) wargaming for training and education. The boundaries between these game types are not hard and fast and are only meant to generally categorize types of gaming. We found service Title 10 games varied widely across these categories by service branch and by event, so we do not put Title 10 games into a separate category. We describe each type in further detail below.

---

[1]   Phillip E. Pournelle, "Designing Wargames for the Analytic Purpose," *Phalanx*, Vol. 50, No. 2, June 2017, p. 50.

Table 2.1
Types of Games

	Goal of Game	
Problem Type	Creative Knowledge	Conveying Knowledge
Unstructured problem	Discovery games	Educational games
Structured problem	Analytical games	Training games

SOURCE: Elizabeth Bartels, "Innovative Education: Gaming—Learning at Play," *ORMS Today*, Vol. 41, No. 4, August 2014.

Wargaming for Concept Development

Marine Corps Combat Development Command/Combat Development and Integration (MCCDC/CD&I) instruction on *concept development* defines a concept as "an expression of how something might be done; a visualization of future operations that describes how warfighters, using military art and science, might employ capabilities to meet future challenges and exploit future opportunities."[2] We use this general definition of concepts when considering wargaming to support concept development.

MCCDC/CD&I acknowledges both unofficial and official concepts, with official concepts being those "formally published by the Service to inform wargaming, modeling and analysis, experimentation, and other capability development activities."[3] Within MCCDC/CD&I, it is official, "validated" concepts that form the basis of capabilities-based assessment and development, personnel, facilities, and policy changes.[4] Types of concepts include the Marine Corps' capstone operating concept, subordinate operating concepts, functional concepts, concepts of operation (CONOPS), and concepts of employment.[5]

Concept development will generally lead capability development. Instruction on the Joint Capabilities Integration and Development System notes that capability requirements should come from service and joint concepts, as well as organizational roles and missions.[6]

---

[2] MCCDC/CD&I, "Concept Development," Quantico, Va.: Marine Corps Combat Development Command/ Combat Development and Integration Instruction 5401.1, February 8, 2016, p. 2.

[3] MCCDC/CD&I, 2016, p. 2.

[4] MCCDC/CD&I, 2016, p. 2.

[5] MCCDC/CD&I, 2016, pp. 2–3.

[6] Joint Chiefs of Staff, *Joint Capabilities Integration and Development System (JCIDS)*, Washington, D.C.: Chairman of the Joint Chiefs of Staff Instruction 3170.01I, January 23, 2015, p. A-3.

Wargaming to Support Capability Development and Analysis

The Marine Corps defines capability development as

> identifying, assessing, validating, and prioritizing Marine Corps capability require-
> ments, gaps, and solutions; identifying acceptable areas to increase, maintain or
> reduce risk across the Marine Corps enterprise; and performing follow-on actions
> that attain the selected and resourced set of materiel and non-materiel capability
> solutions. Capabilities development compromises both capabilities planning and
> solution development.[7]

We consider wargaming to support capability development and analysis as
wargaming to support decisions about capability development or to create products
to support further quantitative analysis of an issue. While analysis need not always
be quantitative, in practice much of the defense gaming that falls within this category
supports operations research (OR) and systems analysis. Activities in this are games
that are more closely related to developing the Program Objective Memorandum, the
final DoD programming process that determines resource allocation.[8] In other words,
these include wargames with the eventual purpose of affecting the decisions in the
defense budget. They also include wargames that take a numbered Support for Strategic Analysis (SSA) scenario and its accompanying Multi-Service Force Development
and further build out events in order to support detailed modeling and simulation
(M&S). Wargames that create a set of operational- or tactical-level events in order to
support service M&S are also included in this category. Organizations whose wargaming activities include wargaming to support analysis include the U.S. Army Center for
Army Analysis (CAA) and The Research and Analysis Center (TRAC).

Science and Technology Wargaming

While it could more generally be considered a part of wargaming to support concept
development, we separate out S&T wargaming into its own category based on our consulting with the sponsor, who identified this as an area of interest. And while there may
be more overlap between S&T wargaming and wargaming to support concept development for a service such as the U.S. Air Force, these two categories might be more
distinct for the Marine Corps. We consider S&T wargames to be games in which the
primary consideration is of S&T concepts and where the insights from a game largely
support S&T development.

Senior Leader Engagement and Strategic Discussion

We consider wargaming and other events that support senior leader engagement and
strategic discussion to be the most distinct category within the types of wargaming

---

[7] Headquarters, U.S. Marine Corps, *Marine Corps Capabilities Based Assessment*, Washington, D.C.: Marine
Corps Order 3900.20, September 27, 2016.

[8] DoD, Under Secretary of Defense (Comptroller), *Financial Management Regulation*, Washington, D.C.: DoD
7000.14-R, June 2017, Vol. 1, p. 15.

that we discuss. This category reflects organizations such as the J-8 Studies, Analysis, and Gaming Division (SAGD) and the Special Operations Command (SOCOM) Wargame Center. In this category the focus is not necessarily on adjudicating outcomes but instead on enabling discussion and gaining feedback from senior decision-makers. While it is the most methodologically dissimilar to the types of adjudicated wargames in the other categories of games, this was also an area in which we could examine approaches and tools, given the WGD's interest in it.

Wargaming to Support Operational Decisions and Plans

Yet another category is wargaming to support operational decisions and plans. This class of games focuses on informing current and future plans and challenges, both at the service level and the joint level. Organizations that conduct this type of wargaming include the CAA, the Air University's LeMay Center for Doctrine Development and Education, the U.S. Army War College (USAWC), and the Center for Naval Analyses (CNA)—usually in support of operational commands.

This type of gaming also appears to be widely done at the service level, outside dedicated wargaming centers. Current service and joint planning processes include COA wargaming as part of planning activities, so wargaming to support operational decisions and plans could theoretically be conducted by any U.S. military organization carrying out planning activities.[9]

Wargaming for Training and Education

To a large extent, the purpose of the original wargame Kriegsspiel was to support training and education within the Prussian Army.[10] Given the organizational purpose of many defense gaming centers and DoD degree-granting institutions to support training and professional military education (PME), this is an important category of defense wargaming. Organizations whose charters put at least some of their gaming into this category include the Army Command and General Staff College (CGSC), the LeMay Center, the National Defense University (NDU), the Naval Postgraduate School (NPS), the Naval War College (NWC), and USAWC.

Within the Marine Corps, MCU is one of the organizations that supports the training and education mission. The Marine Corps War College within MCU uses commercial board games to support classroom instruction, particularly as it relates

---

[9] Joint Chiefs of Staff, *Joint Planning*, Washington, D.C.: Joint Publication JP 5-0, June 16, 2017, pp. V-31–V-32; Headquarters, U.S. Marine Corps, *Marine Corps Planning Process*, Washington, D.C.: Marine Corps Warfighting Publication MCWP 5-10, April 4, 2018, pp. 1–5; Headquarters, U.S. Department of the Army, *The Operations Process*, Washington, D.C.: Army Doctrine Reference Publication ADRP 5-0, May 2012, pp. 2–12; U.S. Department of the Navy, Navy Warfare Development Command, *Navy Planning*, Washington, D.C.: Navy Warfare Publication NWP 5-01, December 2013, pp. 1–5.

[10] Perla, 2012, pp. 35–45.

to strategic thinking.[11] Educational wargaming can benefit from rerunning historical cases in ways that are not always as relevant when wargaming to support future capability development. Nevertheless, best practices and lessons learned from wargaming to support training and education can still inform other categories of wargaming and are an important part of the defense wargaming community.

Time Orientation of Different Wargame Types

As we developed the aforementioned wargame categories during discussions with Marine Corps stakeholders, we noticed that different categories have different time horizons. Figure 2.1 depicts how we view the wargaming categories across these time horizons.

These different time horizons (past, current, and future) in turn can have implications for wargaming. For example, wargames set in the future typically must contend with greater uncertainty and little to no data about a greater range of issues than historical games or games based in the present. This makes adjudication more challenging. Alternatively, training and education games are enabled by a much wider scope of games, including historical games that may not be as applicable to other purposes.

Focus for Future Marine Corps Wargaming Capability

We learned from our discussions with MARCORSYSCOM, OAD, the WGD, and other Marine Corps stakeholders that the priority areas for future WGD capabilities

Figure 2.1
Wargaming Categories and Their Time Horizons

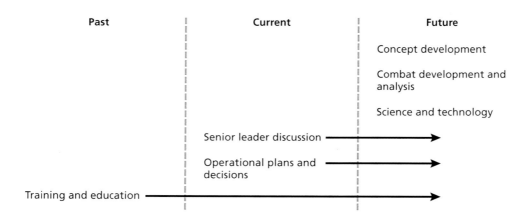

---

[11] James Lacey, "Wargaming in the Classroom: An Odyssey," War on the Rocks, April 19, 2016. The Marine Corps War College is referenced here as using wargaming tools from other PME organizations, such as NDU and the CGSC.

would be wargaming for concept development, capability development and analysis, and senior leader engagement and strategic discussion. S&T wargaming and wargaming to support operational decisions and plans are not expected to be as much of an emphasis, and wargaming for training and education will largely remain outside the WGD's mission. We keep this focus in mind as we make recommendations in Chapter 7.

A Brief History of Wargaming

The history of wargaming is often told as a series of vignettes of specific high-profile games over time. Rather than taking that approach, here we have opted to focus on trends in the application of games. We are also interested in the spread of innovation in gaming: between different countries, different services, different tools, and the military and commercial gaming worlds. In this way we hope to set up the historical roots of key current debates.

This chapter draws from previous attempts to document the application of games to serious military problems. Some of these texts take the perspective of journalists or researchers interested in understanding this peculiar analytical tool.[12] Others are written by members of the community of wargaming practice, documenting their field.[13] While we drew on information and insights from both classes of work, in many ways, this report falls into the latter tradition.

The Early Days of Wargaming

While the use of games, such as chess and go, as a tool to build up strategic thinking is ancient, we typically date the foundations of modern gaming to the Prussian Army staff in the early nineteenth century. In Kriegsspiel games, officers were divided into opposing teams to play out a specific battle, with moves adjudicated based on a set of rigid rules. This represented a critical leap forward in two key dimensions. First, the

---

[12] For notable examples, see Thomas B. Allen, *War Games: The Secret World of the Creators, Players, and Policy Makers Rehearsing World War III Today*, New York: McGraw-Hill, 1987; Igor Mayer, "The Gaming of Policy and the Politics of Gaming: A Review," *Simulation & Gaming*, Vol. 40, No. 6, 2009, pp. 825–862; and Andrew Wilson, *The Bomb and the Computer*, New York: Delacorte Press, 1969.

[13] For notable examples, see Clark C. Abt, *Serious Games*, New York: Viking Press, 1970; Garry D. Brewer and Martin Shubik, *The War Game: A Critique of Military Problem Solving*, Cambridge, Mass.: Harvard University Press, 1979; Matthew B. Caffrey, Jr., *On Wargaming: How Wargames Have Shaped History and How They May Shape the Future*, U.S. Naval War College Newport Papers 43, Washington, D.C.: U.S. Government Publishing Office, 2019; James F. Dunnigan, *Wargames Handbook: How to Play and Design Commercial and Professional Wargames*, 3rd ed., Bloomington, Ind.: iUniverse, 2000; John Thomas Hanley, Jr., *On Wargaming: A Critique of Strategic Operational Gaming*, dissertation, Yale University, 1991; Francis J. McHugh, *Fundamentals of War Gaming*, 3rd ed., Newport, R.I.: U.S. Naval War College, Strategic Research Department, 1966; Perla, 2012; and Martin Shubik, *Games for Society, Business, and War: Towards a Theory of Gaming*, New York: Elsevier, 1975.

game represented realistic terrain and units, and was adjudicated based on rules that took historical attrition and damage into account. Thus, the results of the game could be mapped to the real world far more directly than more abstract games. Second, rather than only being used as a teaching tool, wargames expanded from the classroom to develop approaches intended for use in conflict. As part of their education, Prussian war college students would wargame out likely invasion routes, sharpening their operational plans in competitive play. Students took away the knowledge of strategy and planning, while the staff took away the most successful plan for a range of likely conflicts.

Recognizing the value of competitive, realistic conflict games, the practice of Kriegsspiel expanded to new purposes and actors. From the original focus on specific battles, versions of the game were adopted to look at longer campaigns and wars, as well as small-unit tactical engagements.[14] The use of gaming was expanded from the staff college, as graduates took the approach to the units they commanded as a planning tool. Foreign observers, noting the military success of the Prussian Army, were quick to adopt many of its practices, including gaming. By the turn of the century, the great powers, including England and its empire, France, Italy, Japan, Russia, and the United States, adopted versions of gaming to varying degrees.

Expanding applications in the latter half of the nineteenth century and early years of the twentieth century required the creation of new approaches to conducting the games. For example, increasing demand made it difficult to supply enough adjudicators who could conduct games using the original formal rules. "Free" Kriegsspiel was developed as an alternative whereby experienced veterans would adjudicate outcomes based on their own experience of combat.[15] This proved to be faster and easier than the rule-based approach, and worked well as long as there was a supply of experienced veterans and minimal interference in the results from more senior leaders. Ground rules were refined based on observed unit behavior during conflict, integrating ideas like the will to fight into gameplay. Another example was the development of naval games along similar lines that became a particular focus of British gaming. Similarly, early Italian games added consideration of logistics. These adaptations allowed gaming to influence a broader range of decisions and decisionmakers, making the approach more robust over time.

One other key innovation was the adaptation of military games by civilians as a form of entertainment and education. Many used miniature units, whose movements were played on realistic maps. While some represented famous historical battles, like the Waterloo Campaign (1895), others, like Modern Naval Warfare (1891) took a deep dive into current military capabilities, and invited the user to consider future conflicts. Other games, like Stratego (1906) and H. G. Wells's Little Wars (1913), were some-

---

[14] Milen Vego, "German War Gaming," *Naval War College Review*, Vol. 65, No. 4, Autumn 2012, p. 111.

[15] Perla, 2012, pp. 42–45.

what more abstract in representing forces and terrain but included rules to represent key ideas like unit strength and the fog of war. While access to such games was generally limited to the rich and well educated, they were an important step in expanding the application of games representing military conflict.[16]

Wargaming Before, During, and After World War I

Wargaming was used by both sides to help plan before and during World War I. Prior to the war, gaming was used by both the British and Germans not only to play likely battles and campaigns but also to make broader decisions about force posture. For example, games in 1905 reified the Schlieffen Plan's movement of German forces to the Western Front,[17] while the British opted to reposition forces to enable a faster response to the invasion of Belgium based on wargame results from the same year.[18] At the same time, game findings could just as easily be ignored—Russia encountered many of the same issues it would face in its failed invasion of Germany in early games, but took little action to change its operational plan or force posture to seek a better outcome.[19] These games also showed the weakness of operational gaming—it missed the political context that would drive much of the conflict, and could easily be dismissed by skeptical commanders.

The interwar years saw new approaches to gaming that attempted to bridge these gaps on the part of the Axis powers. The German staff, though numerically limited by the Treaty of Versailles, redoubled its attention to gaming in order to fill key gaps. Operational games allowed the German staff to experiment with equipment like aircraft, tanks, and submarines which it was banned from producing and fielding—effectively, early S&T games. In addition to these operational games, the 1920s saw the use of strategic games to consider the political and strategic implications of military conflict. In addition to considering the political behavior of foreign states, the games also included social and economic considerations, which would prove key to the Germans' approach to prosecuting World War II.[20] Similarly, the Japanese Total War Research Institute gamed out not just operations like the attack on Pearl Harbor but also political considerations.[21] However, these efforts were short-lived, and not uniformly successful. Adolf Hitler halted strategic gaming in Germany in the 1930s. While it was more enduring, Japanese gaming's representation of American decision-

---

[16] Caffrey, 2019, pp. 15–22, 29–31.

[17] Vego, 2012, p. 113.

[18] Wilson, 1969, pp. 28–32.

[19] Wilson, 1969, p 33.

[20] Rudalf M. Hofman, *War Games*, n.p., Historical Division, Headquarters, U.S. Army Europe, 1952.

[21] Perla, 2012, p. 55.

makers assumed far more casualty sensitivity than proved the case, creating misleading results.

The use of gaming by the Allied forces was much more mixed. British and French gaming often discounted technological improvements and underestimated the German threat. While Soviet gaming was limited to formal sand table games that resembled those of the nineteenth century, they appear to have been important in establishing Joseph Stalin's trust in a handful of military officers.[22] The American story is a bit more complicated. While unit-level games declined due to war weariness and limited funds, both the CGSC and the NWC kept gaming at the center of their curricula. Facing limited funding and manpower, leadership sought to make these games relevant to policy leaders, as well as educational for participants. As a result, large portions of the mid- and senior officer corps had shared experiences in wargaming conflicts in Europe and the Pacific. The Army tested out operational approaches and technology like tank destroyers,[23] while the U.S. Navy played a wide range of Japanese capabilities so it would be prepared for a somewhat unknown adversary.[24] At the same time, the Marine Corps used games to work out amphibious operational practices.[25]

The Aftermath of War and the Rise of Operations Research

Perhaps most important, World War II saw the emergence of OR—that is, the use of applied math and engineering to solve industrial-style problems—as a tool to support the war brought about by American and British researchers. OR sections were tasked with identifying and developing new technology such as radar and optimal industrial shipments between the Allies.[26] Following the war, these approaches became increasingly common in defense analysis, benefiting from new discoveries that increased the utilities of computers and from the tenure of Robert McNamara as U.S. Secretary of Defense. As part of this movement, games became increasingly common as a stand-alone tool for research rather than the traditional hybrid of education and staff work.

As OR gained influence, two broad approaches to gaming emerged. The first, which is now closely associated with OR analysis, sought to use the improving computer models to minimize the need for operational experts (that is, military officers) to participate in analysis, leaving the bulk of the work to mathematical professionals.

---

[22] Caffrey, 2019, p. 56.

[23] Christopher R. Gabel, *The U.S. Army GHQ Maneuvers of 1941*, Washington, D.C.: U.S. Army Center of Military History, 1991.

[24] Albert A. Nofi, *To Train the Fleet for War, 1923–1940*, Newport, R.I.: Naval War College Press, 2010; Brian McCue, *Wotan's Workshop: Military Experiments Before the Second World War*, Washington, D.C.: Center for Naval Analyses, 2002.

[25] Caffrey, 2019.

[26] William Thomas, *Rational Action: The Science of Policy in Britain and America, 1940–1960*, Cambridge Mass.: MIT Press, 2015.

This approach aimed to minimize the role of human players by building computers that could take over decisionmaking. The second approach placed far more emphasis on games as a means to elicit information from the players, which could not be appropriately captured without humans in the loop. Of course, most games (and gamers) fell between the two extremes. For example, work by the RAND Corporation in the 1950s and 1960s included "pol-mil" games, often designed to help research teams develop ideas and approaches, as well as the creation of sophisticated computer models.[27]

Both approaches were influential, however. In the McNamara DoD, quantitative modeling and wargames became an important part of the new plans, programs, and budget cycle that determined resources. These types of games also helped shape core strategic documents governing nuclear war, like the Single Integrated Operational Plan. At the same time, senior leaders, including cabinet officers, took part in strategic pol-mil games.

In addition to increasing the styles of gaming executed for military research, the 1950s saw the beginnings of growth in commercial hobby games. In part, this was driven by former service members and civilians having more disposable income and more knowledge of conflict, which raised their interest in the tool. It was also due to the rise of print or board games, which did not require the money and time investment of miniatures. These games covered not just battle-level conflicts on the ground and sea, as earlier games had, but ranges across the domains and levels of war.

Decline and Resurgence in the Post-Vietnam Era

As disillusionment over the Vietnam War grew, so did dissatisfaction with the analytical tools, including gaming, that had influenced the early stages of the war. As a result, gaming experienced a dip in the late 1960s and early 1970s. However, as concerns grew about surging Soviet capabilities, the military schools and other organizations seeking new solutions to the growing conventional and nuclear threat focused anew on gaming's capabilities. In particular, the war colleges began to run integrated games that allowed for exploration of multidomain problems, while also training a new generation of officers who would help to rebuild from the "hollow force."[28]

During these lean years, gaming was sustained by innovation in the commercial board game market. The emergence of SPI as a second major publisher after Avalon Hill, as well as magazines and gaming conventions, created a new, more competitive environment for game design that encouraged innovation. The emergence of both tabletop role-playing games like Dungeons and Dragons (1977) and personal computer (PC)

---

[27] For a summary of early pol-mil gaming at RAND, see Herbert Goldhamer and Hans Speier, *Some Observations on Political Gaming*, Santa Monica, Calif.: RAND Corporation, P-1679-RC, 1959. For a more general overview of gaming, see Milton G. Weiner, *Trends in Military Gaming*, Santa Monica, Calif.: RAND Corporation, P-4173, 1969.

[28] Caffrey, 2019, pp. 87–93.

games in the late 1970s saw the development of an even broader suite of tools. As military gaming sought to rebuild, many professional gamers reached out to the commercial community for inspiration and personnel. This reversed the traditional trajectory from defense gamers to the commercial sector; now the hobby games were supplying the ideas behind a new generation of defense games.[29]

Eventually, gaming in senior-level education began to translate into gaming in the operational and strategic force. One new area was the Navy's Global Games, which began in the late 1970s and set the template for modern Title 10 games. These large future games allowed the service to explore long-term strategic and operational dynamics, and grew over time to involve hundreds of players and staff.[30] By the late 1980s gaming had spread into the combatant commands, which used games to test Operations Plans (OPLANs). To support these efforts, new investments were made in building out wargaming centers (particularly at the PME schools) including computer models to support gaming. These investments continued into the 1980s, and fueled the games of the 1990s.

Post–Cold War Wargaming

Like many other areas of military investment, the work of the 1970s and 1980s was not fully on display until after the fall of the Soviet Union. Wargames helped shape investments in so-called second offset conventional capabilities; now they informed the operations that would validate the power of these conventional capabilities. For example, in advance of Operation Desert Storm, a wide range of games were played that revealed weaknesses in specific COAs, and subsequently shaped the campaign.[31] Wargames also became more integrated in service and the OSD decisionmaking processes, particular in support of the *Quadrennial Defense Review*.[32] For example, growing out of the Navy's Global Games, each service began a Title 10 game series; these were used to inform force structure and acquisition decisions. Many of these efforts were supported by large M&S efforts like the Joint Warfare System and the Joint Simulation System, which aimed to create a joint campaign model that built on previous, service-specific tools. While model-supported campaign analysis was a key element of DoD decisionmaking processes, over time the cost and complexity of the systems limited their utility as a tool to support wargames.

By the beginning of the 2000s these trends had created something of a crisis in gaming. Some games became increasingly large and prone to manipulation in order

---

[29] Caffrey, 2019, pp. 93–94.

[30] Bud Hay and Bob Gile, *Global War Game*, Newport, R.I.: Naval War College Press, 1993.

[31] Caffrey, 2019, pp. 97–103.

[32] Clarence E. Carter, Phillip D. Coker, and Stanley Gorene, *Dynamic Commitment: Wargaming Projected Forces Against the QDR Defense Strategy*, Washington, D.C.: Institute for National Strategic Studies, Strategic Forum, No. 131, November 1997.

to produce desired results. For example, the 2002 Millennium Challenge game for Joint Forces Command sparked controversy when players accused the game staff of manipulating adjudication results to keep the large, expensive game on track.[33] As costs mounted, larger games came under increased pressure to prove their worth, and the programs shifted, started, and stopped as the services and joint community attempted to find a useful balance. Over the next decade, many of the more ambitious M&S efforts would also fall by the wayside due to mounting costs and time delays. As U.S. forces struggled in Afghanistan and Iraq, there was also increased concern that combat models, focused on physical performance of weapon systems, failed to capture the political and social dynamics that were actually driving events on the ground. As major gaming programs faltered, many designers began to refocus efforts on more flexible game designs adapted from commercial board game designs. This effectively created two dueling cultures of gamers: those working on large statutory games or campaign modeling efforts tied to the programs of record and those working on smaller games to support specific studies and programs.

The Utility of Gaming

All things considered, gaming is a broad set of quite flexible tools that have been applied to a wide range of military problems. While games have been used most consistently as an educational tool, they have also seen steady use as an analytical tool, though the popularity and sophistication of analytic gaming has varied over time. As a result, it is worth taking a step back to ask what types of analytical problems games are suited to answer, and which ones they do not.

Fundamentally, games like historical case studies are "small-n" types of tools, which examine one or a couple of games rather than behavior over repeated experimentation. This means that games are well suited to observing rich contextualized data about group decisionmaking processes. The other primary characteristic is that games, like all M&S, are artificial environments—they can explore events that have not happened. However, to be manageable and useful, games must take away much of the true complexity of the world. Thus, games are a way of gaining rich, contextual data from inside a simplified model of the world.

Most designers are uncomfortable with claims that games are predictive of specific system performance or personal performance because the small number of iterations in an artificial environment is exceedingly unlikely to generate reliable point predictions. As a result, there is broad discomfort with using games in experiment

---

[33] Micah Zenko, "Millennium Challenge: The Real Story of a Corrupted Military Exercise and Its Legacy," *War on the Rocks*, November 5, 2015.

designs (though the verdict on games as quasi-experiments is still out),[34] or with any claim that games in some way "prove" results rather than provide insights about decision contexts, behaviors, and outcomes that can be further examined in future studies.

On the other hand, using games as a tool to integrate expert mental models and foster creativity is much more broadly accepted.[35] In large part, designers believe this is due to games creating a "story-living experience" that enables participants to internalize their experiences.[36] That is, the vivid artificial environment of a game lets players explore different decisions in a way that feels real at the time—providing, in effect, artificial experience. As a result, games make it possible to walk through situations, debates, and plans and to discover findings that, if not truly novel, may have impact and relevance that they would not in another context. This is a mixed blessing, as games that incorrectly specify the decisionmaking environment can be "seductively" wrong and encourage participants to believe in findings more than might be warranted. However, this concern does not remove the very real power of games to bring to the surface insights that, once stated, are perceived as obviously true by participants and observers alike.[37]

---

[34] Edward A. Parson, "What Can You Learn From a Game?" in Richard Zeckhauser, Ralph L. Keeney, and James K. Sebenius, eds., *Wise Choices: Games, Decisions, and Negotiations*, Boston, Mass.: Harvard Business School Press, 1996; Margaret Polski, "Are Wargames Quasi-Experiments?" panel discussion at the 83rd MORS Annual Symposium, Alexandria, Va., June 24, 2015.

[35] Parson, 1996.

[36] Peter Perla and E. D. McGrady, "Why Wargaming Works," *Naval War College Review*, Vol. 64, No. 1, Summer 2011.

[37] Robert A. Levine, Thomas Schelling, and William M. Jones, *Crisis Games 27 Years Later: Plus C'est Déjà vu*, Santa Monica, Calif.: RAND Corporation, P-7719, 1991.

CHAPTER THREE

The Current Environment for Wargaming

The current Marine Corps interest in wargaming is taking place in the context of several trends, which we discuss in this chapter. One trend, perhaps the most visible, is the renewed push across DoD to increase and improve wargaming. We see this in the form of top-down guidance from senior leadership on a number of fronts and in the variety of wargaming activities sponsored by DoD leadership. Another has been the bottom-up developments that have been ongoing within the wargaming community for several years—in particular, developments in expanding communities of practice (CoPs) and formalizing wargame education. Although the recent DoD-wide emphasis on wargaming has increased interest in wargaming education, the wargaming community has for some time been attempting to improve practice and knowledge within the field.

Beyond wargaming developments in defense circles, the current resurgence in wargaming also takes place against a backdrop of developments in commercial gaming. While it is not as directly tied to defense wargaming practice, we also briefly cover this in Chapter 3.

Senior Leader Emphasis

A series of recent senior leader memos and other expressions of interest in wargaming are what is behind the current, renewed DoD interest in wargaming. A series of communications, from the U.S. Secretary of Defense on down, have been part of an effort on the part of DoD leadership to increase wargaming activity. DoD leaders have also explicitly tied wargaming with innovation, indicating a deeper desire to spur change and new ways of approaching challenges during a time of perceived technological and geopolitical change.

Secretary of Defense Chuck Hagel issued a November 2014 memorandum titled "The Defense Innovation Initiative," stating the need to "pursue innovative ways to sustain and advance military superiority in the 21st Century." He specifically argued that a "reinvigorated wargaming effort will develop and test alternative ways of achieving our strategic objectives and help us think more clearly about the future security environment."[1] Then–Deputy Secretary of Defense Robert Work issued a memorandum in February 2015 titled "Wargaming and Innovation" in which he argued that wargames can "spur innovation and provide a mechanism for addressing emerging challenges, exploiting new technologies, and shaping the future security environment."[2] In 2015 Deputy Secretary of Defense Robert Work and Vice Chairman of the Joint Chiefs of Staff General Paul Selva also published "Revitalizing Wargaming Is Necessary to Be Prepared for Future Wars," in which they articulated why they believe wargaming can support innovation if better integrated into programmatic actions. This article also laid out some of the initiatives they had sponsored to further wargaming's development.[3] These initiatives included a wargaming repository overseen by OSD Cost Assessment and Program Evaluation (CAPE), the formation of the Defense Wargaming Alignment Group (DWAG), and better inclusion of allies within DoD wargaming efforts.[4] The quad-chairs of DWAG are the Joint Staff, OSD CAPE, the OSD Office of Net Assessment (ONA), and the OSD Office for Policy.[5]

This discussion on wargaming from the highest levels of the Pentagon was soon followed by service-level guidance within the Department of the Navy. Then–Secretary of the Navy Ray Mabus issued a May 2015 memorandum that identified wargaming as "an invaluable method used to assess new ideas, question existing practices, stimulate innovation, and develop new operational concepts in a risk-friendly environment."[6] The memorandum outlined actions for the Marine Corps and the Navy to reinvigorate their wargaming capabilities and institutions, and to expand wargaming's influence on operations and the Planning, Programming, Budgeting, and Execution System (PPBES) process. Another activity directed by the Navy has been the creation of a naval wargaming virtual CoP.[7]

Other Departments Activities

In 2016 DWAG sponsored a wargaming special meeting in collaboration with the Military Operations Research Society (MORS), which was also open to allies and the

---

[1]   Chuck Hagel, U.S. Secretary of Defense, "The Defense Innovation Initiative," memorandum, Washington, D.C., November 15, 2014.

[2]   Robert Work, U.S. Deputy Secretary of Defense, "Wargaming and Innovation," memorandum, Washington, D.C., February 9, 2015.

[3]   Bob Work and Paul Selva, "Revitalizing Wargaming Is Necessary to Be Prepared for Future Wars," War on the Rocks, December 8, 2015.

[4]   Work and Selva, 2015.

[5]   Mark Gorak, "Introduction," special issue, "Modeling and Simulation Special Edition: Wargaming," *CSIAC Journal*, Vol. 4, No. 3, November 20, 2016, pp. 5–7.

[6]   Ray Maybus, U.S. Secretary of the Navy, "Wargaming," memorandum, Washington, D.C., May 5, 2015.

[7]   M. Brian Ross, U.S. Naval War College, "Naval Wargaming Becomes a Collaborative Community," press release, Washington, D.C., June 2017.

rest of the U.S. government. The purpose was to cultivate and expand the pool of wargamers within the DoD. The 2016 special meeting emphasized learning by doing, and had tracks that included adjudication, game design, terms of reference, matrix-style games, and wargame analysis.[8] MORS conducted another wargaming special meeting in October 2017, again with DWAG sponsorship.[9] The Office of the Under Secretary of Defense for Acquisition and Sustainment, Cyber Programs, hosted the October 2018 MORS cyberspace special meeting on wargaming and analytics.[10]

DoD has also taken other steps to increase wargaming skills. For example, through the OSD Wargaming Education Initiative, the NPS is seeking to develop a faster and lower-cost automated education and assessment system that can provide wargaming skills to DoD and to allied and partner organizations.[11]

At the service level, the Marine Corps will build a new wargaming center in Quantico, Virginia. The new center is expected to expand the number of games that the Marine Corps currently runs.[12]

Developments Within the Wargaming Community

Although key senior leaders clearly drove the renewed DoD-level interest in wargaming, this greater focus happened against a backdrop of continuing, bottom-up activity by the DoD and allied professional wargaming communities. In particular, the wargaming CoPs and efforts by wargamers to offer more formal instruction on gaming have been important bottom-up activities.

Defining Communities of Practice

We begin with a brief introduction to CoPs. A CoP is a group of people with a shared interest in an activity, who interact regularly to learn and improve the way they conduct that activity. There is the shared domain of interest, the community, and the practice.[13] For example, a group of photocopier repairmen who share information on

---

[8] Phillip Pournelle, ed., *MORS Wargaming Special Meeting, October 2016: Final Report*, Alexandria, Va.: Military Operations Research Society, 2017, pp. 3–10.

[9] Phillip Pournelle and Holly Deaton, eds., *MORS Wargaming III Special Meeting, 17–19 October 2017: Final Report*, Alexandria, Va.: Military Operations Research Society, April 2018, p. 2.

[10] MORS, *MORS Special Meeting: Cyberspace Special Meeting—Wargaming & Analytics, Terms of Reference*, Arlington, Va.: Military Operations Research Society, October 17, 2018, pp. 1–3.

[11] Jeff Appleget, Fred Cameron, and Robert E. Burks, "Wargaming at the Naval Postgraduate School," special issue, "Modeling and Simulation Special Edition: Wargaming," *CSIAC Journal*, Vol. 4, No. 3, November 2016, pp. 18–23.

[12] Todd South, "Marine Wargaming Center Will Help Plan for Future Combat," *Marine Corps Times*, September 19, 2017.

[13] Etienne Wenger, *Communities of Practice: A Brief Introduction*, paper presented to the STEP Leadership Workshop, University of Oregon, October 2011, pp. 1–2.

repairing copiers is a CoP. An active practice or skill distinguishes CoPs from communities of interest, who may meet and have shared interests but who do not work to transmit knowledge to each other on how to improve their skill at an activity. Both corporations and government entities often encourage CoPs and see them as a key component to managing knowledge within an organization.[14] CoPs are particularly important where more formalized means of transmitting knowledge are lacking.

CoPs can be best understood as a convergence of three elements: mutual engagement, joint enterprise, and shared repertoire.[15] First, membership in a group inherently requires mutual engagement. This mutual engagement and membership define the community, whether it is professional, recreational, or political. This mutual engagement introduces relationships, social complexity, and diversity.[16] Second, a joint enterprise provides coherence to the CoP—in this case, the practice of wargaming. By constantly negotiating the goals and means of the enterprise, the members of a CoP create a sense of purpose and accountability within the group.[17] Third, a shared repertoire in the form of specific terms, stories, tools, or a common history reinforces the relationships within the CoP. This is critical to the creation of a group identity that bolsters both its members' engagement with each other and the collective joint enterprise.[18] All these aspects reinforce and intertwine with each other to form the basic dynamic of a CoP.

Wargaming Communities of Practice

CoPs have played an important role in creating and sharing knowledge about defense wargaming. There are three overlapping CoPs centered around wargaming within the wider defense community: the MORS Wargaming CoP, Connections, and the international defense wargaming CoPs. There is significant overlap among these communities, and Connections is the longest running.

Officially incorporated in 1966, MORS is a professional society that itself is a CoP for OR and systems analysts engaged in national security work—largely for DoD and the Department of Homeland Security.[19] MORS's annual conference is organized into standing workings groups, including one focused on gaming. The organization also includes standing CoPs that meet with more frequency and host workshops outside the annual conference format. The MORS Wargaming CoP was established in

---

[14] Wenger, 2011, p. 4.

[15] Etienne Wenger, *Communities of Practice: Learning, Meaning, and Identity,* New York: Cambridge University Press, 1998, p. 73.

[16] Wenger, 1998, pp. 73–76.

[17] Wenger, 1998, pp. 77–81.

[18] Wenger, 1998, pp. 82–85.

[19] MORS, "About MORS," webpage, undated a.

2008 and is one of the more active MORS CoPs.[20] As the majority of MORS members come from the defense community, the MORS Wargaming CoP arose from the desire to further the theory of wargaming and its applications within the national security field, particularly within the DoD.[21] Originally grouped with M&S, wargaming eventually became a separate area and its own CoP.[22] The MORS Wargaming CoP focuses on developing best practices for the definition, design, execution, and analysis of wargames. This involves designing and executing a formal wargame certificate program; sponsoring workshops, lecture series, and game demonstrations; and other activities in support of the DoD wargaming community.[23]

In contrast, Connections largely attracts a mixture of gamers from the DoD community and commercial and professional gamers. At the direction of Col John Warden, then-commandant of the Air Command and Staff College (ACSC), Matt Caffrey initiated the Connections Wargaming Conference shortly after the First Gulf War. Initially started as an annual conference, Connections gradually coalesced into the first multidisciplinary wargaming CoP. From its inception, Connections sought to serve as a venue where gamers from academia, the commercial sector, the government, and the military could leverage and combine their expertise.[24] The MORS Wargaming CoP and the Connections CoP have substantial membership overlap, and are generally seen as complementary efforts to build bridges between different gaming centers that might not otherwise have venues to exchange practices.

There are also international wargaming CoPs that have overlap with the U.S. defense wargaming community. In recent years, affiliated Connections wargaming conferences have been founded to serve international audiences, including the United Kingdom in 2013 and Australia and the Netherlands in 2014.[25] As a result, the Connections CoP has provided a shared experience base, lexicon, and vision, which in turn allow members to share and collectively improve their knowledge of gaming. The King's College London Wargaming Network, launched in December 2018,[26] also overlaps with the U.S. and overseas wargaming CoPs.[27]

---

[20] MORS, "Communities of Practice," webpage, undated b. This study's lead author was previously an adviser to the MORS CoP.

[21] Email interview with MORS wargaming CoP member, June 29, 2017a.

[22] Email interview with MORS wargaming CoP member, June 29, 2017a.

[23] Email interview with MORS wargaming CoP member, June 29, 2017a.

[24] Email interview with U.S. Navy wargaming CoP member, July 17, 2017.

[25] Connections UK, "Aim and Purpose," webpage, undated; Connections Wargaming Conference, homepage, undated.

[26] King's College London News Centre, "New Wargaming Network Launched at King's," press release, London, December 3, 2018.

[27] While the Navy attempted to stand up a virtual wargaming CoP in 2017, there appeared to be little activity in this space as of April 2019. Email interview with U.S. Navy wargaming CoP member, July 17, 2017; email interview with U.S. Navy wargaming CoP member, April 24, 2019.

Wargaming Education

Wargaming has largely been a field with limited available education, whether through professional development programs or formal educational institutions.[28] The majority of wargaming organizations have historically focused on providing wargaming in support of analysis or utilized it as an educational tool—not as a topic of education itself. As a result, wargaming does not have a centralized educational institution or academic vision of how its theory, principles, and application are tied together.

However, there has been movement within the wargaming community to develop formal education on wargaming. Several programs have emerged in the past few years across the defense enterprise whose explicit goal is educating the next generation of wargamers and advancing the field of wargaming. Within the military, Air Force Materiel Command (AFMC), the NPS, and the NWC offer wargaming classes. MORS has also instituted a professional certification program in wargaming in an effort to teach and document skills. There is also a growing contingent of wargame practitioners who advocate for the establishment of a formal wargaming degree program within an academic institution. The belief is that such a program will cultivate a wider and diverse generation of young wargamers, as well as developing additional formal theory and method where there is currently mostly practice. Furthermore, institutionalized wargaming education outside the defense community may create a locus for innovation and institutional knowledge in the face of cycles of gaming practice popularity within the defense practitioner community.

Trends in Commercial Gaming

Finally, although it is beyond the scope of this report to do full justice to the trends in commercial gaming, we note a few of them here. The commercial gaming world is another influence in the overall context for national security and defense wargaming, both demonstrating new ideas and shaping expectations about gaming in general.

Most important, more people are playing more games, giving both designers and players of DoD games a greater library of games to inspire professional work. The digital game market, which includes PC, console, and mobile games, is substantially larger than the tabletop game market, which includes board, card, dice, and miniature games. However, both have seen meaningful growth over the last decade. Industry watchers estimate that the size of the global board game market grew from $9.3 billion in 2013 to $9.6 billion in 2016, a 3 percent increase. The U.S. market grew 28 percent from 2015 to 2016.[29] For comparison, the global market for digital games was

---

[28] Email interview with MORS wargaming CoP member, July 6, 2017b.

[29] Christine Birkner, "From Monopoly to Exploding Kittens, Board Games are Making a Comeback," *Adweek*, April 3, 2017.

estimated to be worth $99.6 billion in 2016, up 8.5 percent over the previous year. U.S. revenues were estimated at $23.6 billion, with 4.1 percent growth.[30]

Digital game platforms like the Apple and Google app stores and Steam, online retailers like Amazon, and crowdfunding vehicles like Kickstarter have made it easier for consumers to access a wider range of tabletop and digital games. As a result of this proliferation of games, there are many more models for game designers to leverage in their professional gaming work. For example, in interviews at gaming centers we saw a wide range of games that had been adopted, particularly to support DoD educational gaming. At the same time, greater exposure to games is changing the expectations of players. For example, we encountered a frequent complaint that players expected DoD digital game graphics to be as good as major commercial titles like the *Call of Duty* series (2003–2016). These expectations are hard to meet given the far smaller budget of DoD gaming acquisition, compared with games with fan bases in the millions of players.

Another trend that we note is that more gaming now incorporates social interaction, which can make gaming more appropriate to the purposes of professional DoD games. In the digital game space this trend is visible even in causal or story-focused digital games which did not previously emphasize social dynamics. Causal games like *Farmville* (2009) or *Words with Friends* (2009) now include many more options for social interaction than did previous generations of games, like *Minesweeper*, or digital crossword puzzles. Similarly, easier and more engaging multiplayer modes have begun to eclipse single-player campaign-style games. In games ranging from free-to-play mobile games like *Clash Royale* (2016) to console games like *Halo* (2001–2016), more attention is given to multiplayer features.[31] In the tabletop gaming industry, the rise of "Euro-" or "German-style" games like *Settlers of Catan* (1995) puts a similar focus on accessible games that reward interaction between players.[32] This emphasis on social interaction can make these games particularly relevant to designers of DoD games that seek to build relationships or synthesize knowledge from different players. The focus on accessible mechanics can also make these games more useful to designers who need to be able to engage professional audiences that may not be very familiar or comfortable with more complicated mechanics.

Closely related to this trend is a focus on increasing the playability of games while maintaining high levels of detail and dynamic gameplay. In the past, one of the key dilemmas of manual-style games was the inverse relationship between complexity and playability. As the level of detail increases in a game, rules typically grow increasingly complex, ultimately reducing playability. Many games from the "golden age" of 1970s

---

30 Newzoo Games, *Free 2016 Global Games Market Report: An Overview of Trends & Insights*, San Francisco: Newzoo Games, June 2016.

31 Interview with video game industry analyst, Los Angeles, September 4, 2016.

32 Dan Jolin, "The Rise and Rise of Tabletop Gaming," *Guardian*, September 25, 2016.

hobby gaming required hours merely to read the rules—a trend taken to parody in *Campaign for North Africa* (1979). Such games were highly accurate, but required players to learn complicated rules that included many exceptions. These were difficult to track even for experienced players. In response, designers began to experiment with different presentations of game rules to make play more intuitive. This approach is a hallmark of the Eurogame genre, but has spread more broadly. For example, card games such as Magic: The Gathering (1993) and Dominion (2009) print the specific rules for a given card on the card itself, removing the need to look up exceptions in a long reference book. At Connections 2017, a commercial developer offered another example, whereby instead of having a rule in which movement of individual pieces were modified when traveling over certain terrain, the designer would just increase the number of spaces in accordance with the difficulty of the landscape. Thus, terrain remained a factor without introducing cumbersome rules. Commercial developers argue this will help manual games achieve higher levels of complexity while simultaneously enhancing playability.

CHAPTER FOUR

The Marine Corps' Vision for Next-Generation Wargaming

Given the current state of gaming, as described in Chapters 2 and 3, we can turn to considering the WGD's vision for the future of wargaming. This chapter reflects our discussions with the WGD in 2016–2017. The Marine Corps envisions NGW as the next evolution of wargaming, combining time-tested best practices and emerging technology to expand gaming possibilities. Fundamentally, the vision is rooted in the belief that game design has often reflected trends in hobby gaming: first from manual tabletop games, then from games with computerized adjudication models. This raises a question: How can defense games today embrace more characteristics of the current generation of commercial wargaming? The Marine Corps NGW vision offers one approach to incorporating tools and approaches to gaming that will be familiar to new generations of Marines.

Generational Constructs of Wargaming

In both the WGD's concept of NGW and in what the study authors have observed in interactions with members of the wargaming community, there appear to be generational differences in how game designers and participants conceive of wargaming. In this section we flesh out the idea of generational constructs and highlight the ways that the different experiences with civilian gaming technology may have shaped expectations about defense wargaming.

A key concept in NGW is that each generation of gamers possesses a unique gaming paradigm, largely shaped by the games that generation played in its adolescence. This report does not pass judgment on whether different generational constructs are better or worse. We simply propose that different generations within the wargaming community default to and reference the most familiar, dominant types of commercial games that were available during gamers adolescence and young adulthood. We argue that these unconsciously internalized archetypes can affect what people view as legitimate game design, mechanics, interactivity, immersion, gameplay, game technology, and speed of play. In essence, the games played during adolescence fundamentally shape a player's paradigm of gaming—in terms of gameplay, style, adjudication, and immersion. This may even underlie certain debates within the wargaming community about what constitutes a good user experience and what is even possible in gaming. Notably, there are also potential manpower implications because of the different skill sets that wargamers acquired through gaming in their youth. We acknowledge that any discussion of generational trends is by its very nature very broad and emphasize that these are very general observations.

We begin with the baby boomers, who currently represent the most experienced and senior wargamers in defense and national security wargaming. We argue that for this generation, the primary commercial game experience was largely with social board games and, to a lesser extent, miniatures games. This is clear from the intuitive understanding that many in this generation of gamers bring to professional gaming: They understand the rules, conventions, terminology, and deeper patterns of hex games and other manual tabletop wargames with little or no additional explanation. Their personal gaming experiences inherently affect their professional wargaming—inverting the direction of diffusion, which until the 1970s had tended to flow from defense gaming to hobbyists rather than the other way around. Consequently, this generation of gamers has mastered and proliferated manual-style games to the point of dominance within the DoD community. This mastery of the board game literature and the range of mechanics that could be applied to other problem sets is not trivial, and it constitutes a specialized expertise that takes years if not decades to build.

Among the wargamers we interviewed for this report, the applicability of their commercial gaming experience is clear. These gamers praise the diversity of problems that have been tackled in the commercial space and their resulting ability to adopt new, but well-tested, approaches to tackling a wide range of issues. They also praise the transparency of board game mechanics: Designers, players, and analysts can easily observe the way in which the game translated player actions into outcomes, allowing deeper engagement with the causal arguments underpinning the game. Finally, this generation of gamers has found eager collaborators in the commercial board game industry, with designers like Larry Bond, James Dunnigan, Mark Hermann, and Joe Maranda producing prolific work in both spaces. We note, however, that little of the baby boomer experience in commercial gaming is digital.

Generation X and millennial gamers share more of a common commercial gaming background involving PC, console, online, and social media games. Both came of age on PC and console games, though to different degrees of sophistication. Generation X gamers grew up on Atari and Nintendo consoles and arcade games. Popular Gen X console titles include *Space Invaders* (1978), *Super Mario Bros.* (1983), and *The Legend of Zelda* (1986). Gen X gamers also saw the beginning of early handheld games with the popular Nintendo Gameboy. However, what stands out about Gen X gaming is its stand-alone nature; this generation's gaming experience during its formative years was less social than those of generations before or after.

Millennial gamers played more technologically advanced consoles like Xbox and PlayStation, as well as more sophisticated PC games that increasingly leveraged the connectivity of the internet. Leveraging millennial familiarity with gaming consoles, the U.S. Army outfitted actual ground robots and unmanned aerial systems with Xbox controllers a decade ago.[1] Furthermore, unlike previous generations, millennials played games on an unprecedented scale in real time, best characterized by massively multi-player online games. Examples of such games include *Everquest* (1999) and *World of Warcraft* (2004). Also important was the rise of multiplayer game modes that made strategy games like *Star Craft* (1998) more engaging than tradition single-player computer and console games. Overall, the gaming experiences of Generation Xers and millennials translate most directly into the training games and simulators utilized by the military today, such as *UrbanSim* (2006) and the *Virtual Battlespace* (2015) series of theater operations simulators.[2] In fact, many of the tabletop games favored by professional baby boomer wargamers are unfamiliar terrain to Gen Xers and millennials.

The latest generation, known as Generation Z, iGen, or digital natives, is currently forming its own generational construct of gaming. This emerging gaming construct will likely be characterized by the advent of mobile and networked gaming systems, increasingly featuring augmented reality (AR) and virtual reality (VR). Examples include *Castle Battles* (2016), *Pokémon Go* (2016), and *Planescape: Torment* (2017). The advent in 2017 of the Nintendo Switch introduced Gen Z to a hybrid platform that was both a portable handheld and networked home console.[3] That year also marked the launch of the immensely popular online game *Fortnite: Battle Royal* from Epic Games.[4] At present, the implications of the gaming experiences of Gen Z for professional wargaming are unknown. However, this report argues that Gen Z, like previous generations, will indelibly shape the wargaming field in the coming years. The effects of Gen Zers and younger millennials are already manifesting themselves in the research and development of AR and VR for national security gaming applications, which has begun to be explored by organizations including the CNA, the Intelligence Advanced Research Projects Activity, the Naval Undersea Warfare Center (NUWC), and RAND.[5]

Table 4.1 summarizes the different commercial gaming experiences and their applications to professional wargaming for the generations we have discussed. Again,

---

[1]   David Hambling, "Game Controllers Driving Drones, Nukes," *Wired*, July 19, 2008.

[2]   A robust description of these efforts is offered in Corey Meas, *War Play: Video Games and the Future of Armed Conflict*, New York: Houghton Mifflin Harcourt, 2013.

[3]   Kevin Webb, "How Nintendo's Handheld Video Game Consoles Have Evolved over the Past 30 Years, from the Original Game Boy to the Switch," *Business Insider*, April 23, 2019.

[4]   Colin Campbell, "How Fortnite's Success Led to Months of Intense Crunch at Epic Games," *Polygon*, April 23, 2019.

[5]   Intelligence Advanced Research Projects Activity, "Using Alternate Reality Environments to Help Enrich Research Efforts (UAREHERE)," IARPA-RFI-13-03, Washington, D.C., March 11, 2013.

Table 4.1
Generational Constructs of Gaming

Generation	Commercial Games	Wargame Application
Baby boomers	• Social boardgames • Miniatures games	Tabletop (manual) games
Generation X, millennials	• Stand-alone PC/console • Multiplayer console • Online multiplayer • Social media	Training games
Generation Z	• Mobile, digital, networked • AR and VR	Unknown

the point is not that any generational construct is better or worse, but simply that different generations begin with different reference points about the nature of gaming and varying levels of familiarity with different gaming techniques and technology.

Ultimately, NGW aims to incorporate emerging aspects of gaming among millennial and Gen Z gamers to evolve the current paradigm of wargaming—both in terms of technology and methodology. This is not an argument that the simple integration of technology inherently creates better games. Technology, no matter how advanced, is no substitute for sound fundamental gaming principles. The basic principles of gaming remain time-tested in their efficiency and utility. Nevertheless, as elaborated in the following section, NGW strives to enhance and build upon traditional gaming methods and techniques to further evolve wargaming as a whole. This report argues that Gen Zers and millennials, like previous generations, will indelibly shape the wargaming field in the coming years.

The 2017 Vision for Next-Generation Wargaming

In this next section we describe the WGD's concept for NGW as presented to wargaming audiences in 2017. We do this in order to document the perspective of an important stakeholder in Marine Corps wargaming just before the Marine Corps undertakes a considerable expansion of its wargaming capability. This is the context in which we make our recommendations. Discussions about what the next generation of defense wargaming could or should look like also raise important questions about technology, adjudication, wargaming personnel skill sets, user experience, and stakeholder expectations.

The WGD's vision for NGW can best be characterized by four aspects: continuous play without turns, real-time in-stride adjudication, continuously evolving scenarios, and an emphasis on immersion.[6] These characteristics are designed to enable fluid and adaptive decisionmaking by players while simultaneously providing adjudication rooted in pragmatic realities and data. At the 2017 Connections Wargaming Conference, the WGD described NGW as "a wargaming art and method that will seamlessly represent an evolving operational environment and accommodate the agility, imagination, and speed of innovative thought."[7] Through NGW the Marine Corps seeks to establish itself as a premier wargaming center—in terms of technology, capabilities, processes, and methods.

With "in-stride adjudication" or "no-turn" gaming, the artificiality of turn-taking in wargaming is eliminated or at the very least mitigated. Players make simultaneous actions without perfect knowledge of gameplay. This incentivizes players to make faster decisions with less information. This simulates a sense of fog and friction within gameplay while rewarding speed of action and the rapid processing of information.[8] However, this requires an absence of rigid steps in gameplay and disaggregated adjudication, which past wargaming methods have failed to achieve in a realistic and timely fashion. To achieve this level of rapid-pace decisionmaking and adjudication, NWG will require consistent and near-instantaneous adjudication, most likely relying on a computer simulation–based model or artificial intelligence.[9] Such no-turn games are beginning to be used in operational wargames for the DoD, but are still an area of exploration for game designers without a strong set of shared best practices.[10]

As depicted in Figure 4.1, NGW seeks to leverage a collaborative gaming engine where blue and red cells work separately to present their respective COAs to the white cell, where adjudication will occur. Then scenario updates and adjudication results will be digitally transmitted to players via their respective touch tables. Ideally, the collaborative gaming engine will provide seamless in-stride adjudication to blue and red cell actions. Afterward, the gameplay information would be transmitted to an interactive device in the wargaming arena, labeled Ellis Hall in Figure 4.1. Within the wargaming arena, players will be able to assess decision anatomy in a plenary discussion and be led in guided hotwash.[11]

---

[6]   Interview with members of the WGD, Quantico, Va., March 23, 2017.

[7]   WGD, Connections Wargaming Conference 2017 briefing, Quantico, Va., August 2, 2017.

[8]   Interview with members of the OAD, Quantico, Va., March 23, 2017; interview with members of the WGD, Quantico, Va., March 23, 2017.

[9]   Currently, there are a handful of games that represent this kind of no-level gaming such as the Air War College's Pegasus game, McGill University's Brynenia, and the United Kingdom's Megagames.

[10] For a more recent treatment of in-stride adjudication in wargaming, see Merle Robinson, Stephen Downes-Martin, and Connections US Wargaming Conference 2018 Working Group, *In-Stride Adjudication*, Washington, D.C.: NDU, July 19, 2018.

[11] WGD, 2017.

Figure 4.1
The Collaborative Gaming Engine

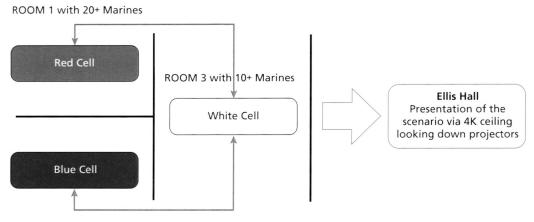

SOURCE: WGD, 2017.

In the past, a wargame's overarching scenario was constrained by a degree of predetermination, typically the result of game scope and game mechanics. A manual wargame cannot feasibly contend with the plethora of uncertainties and variance in COAs. Even computer-aided adjudication possesses its limits in enabling a truly evolving scenario within a wargame. However, through continuous play and in-stride adjudication, one aim is to incorporate the decisions of players in order to present an adaptive set of challenges and evolving scenarios. Ideally, NGW will be a dynamic system in which variables that shape risk, friction, player actions, and their subsequent consequences combine to propel the wargame's scenario forward.[12] Player autonomy is expected to have a greater effect upon the wargame's outcome.

Finally, NGW seeks to set an "unprecedented" level of immersion within its wargames, moving away from wargaming's dominant board game–style format. Typically, a strong immersive experience is a hallmark of educational wargames in which the priority is to train and educate participants. However, with advances in technology in terms of commercial video games and M&S in the military, demand for a greater immersive experience in *all* wargames has increased. Consequently, NGW strives to emulate commercial gaming, putting an emphasis on player engagement with a dynamic playing interface. Although tentative, the Marine Corps seeks to leverage emerging technologies such as AR, VR, and gaming networks to construct an immersive gaming environment.[13] Naturally, this approach requires substan-

---

[12] WGD, 2017.

[13] Interview with members of the OAD, Quantico, Va., March 23, 2017; interview with members of the WGD, Quantico, Va., March 23, 2017.

tial investment in state-of-the-art technology to provide a truly immersive playing environment within wargames. Although the specific technologies required are not specified, the NGW's methods, models, and tools can be best characterized by the following:[14]

- **decision tools,** capable of moving, displaying, synthesizing, and manipulating information so that the anatomy of a decision can be understood and an autopsy of the results can be assessed
- **scenario tools,** capable of representing the effects of the decisions and events by evolving to accommodate the consequences and depicting the resultant situation
- **adjudication tools,** capable of in-stride resolution of conflict and the management of results
- **synthesis tools,** which facilitate visualization and collaboration by allowing inputs from different sources to be merged, displayed, and manipulated.

Despite its reliance on technology to realize its vision, the WGD argues that "NGW is not about technology but about facilitating, synthesizing, and assessing the process of human decisionmaking which drives the wargame."[15] The Marine Corps understands advanced technology is not a substitute for sound gaming principles and design. NGW is envisioned as a convergence of emerging technologies (e.g., scenario, decision, collaboration, adjudication tools) and time-tested wargaming theory and principles—a feat M&S integration in manual wargames has yet to achieve.

Guidance from the deputy commandant of MCCDC/CD&I identified the core function of the Marine Corps' future wargaming capability as facilitating and communicating the validation or invalidation of service concepts in order to support force development, capability development, and requirements generation.[16] As such, the requirements for NGW and its planned wargaming center include, but are not limited, to the following:[17]

- The ability to assess "future operating and functional concepts, in order to establish and define detailed future capability requirements, across the pillars of doctrine, organization, training, materiel, leader development, personnel, facilities, and policy (DOTMLPF-P) in terms of tasks, conditions, and standards."

---

[14] WGD, 2017.

[15] WGD, 2017.

[16] U.S. Marine Corps, Combat Development and Integration, *Initial Capability Document for Marine Corps Wargaming Capability,* version 1.3, Quantico, Va., May 2017, p. 1.

[17] U.S. Marine Corps, *Initial Capability Document for Marine Corps Wargaming Capability,* May 2017, p. 4. For more current information on specific decisions on the new Marine Corps wargaming center, see John Maurer, "USMC Wargaming Capability," PowerPoint presentation, Marine Corps Systems Command, Quantico, Va., January 24, 2019.

- The rapid testing of current war plans in a coalition environment, through examination of existing OPLANS and CONOPS.
- "Simulat[ing] employment of the full range of friendly capabilities and accurately and realistically depict projected threat order of battle and capabilities."
- "Conduct[ing] analysis of wargame-derived date to determine the nature and the extent of capability gaps, possible solutions, prioritization of capabilities, gaps, and solutions and balancing risk in resourcing."

The WGD's vision of NGW seeks to combine the best of technology, analytical rigor, and innovative methods to push the boundaries of wargaming. However, the WGD also noted that it was only at the beginning of realizing its vision for NGW. Nevertheless, despite its infancy, NGW provides critical context for the ambitions of the Marine Corps for its wargaming capability. Our interviews at wargaming centers were conducted with NGW in mind. Subsequently, the recommendations made in Chapter 7 were specifically designed to facilitate the Marine Corps to build and execute the NGW concept.

CHAPTER FIVE

Wargaming Centers

Overview

We were able to interview members of a number of wargaming organizations in our previous effort for the wargaming tools and approaches catalog, and it is these interviews that form our framework for wargaming tasks, COAs, and recommendations in this study. Because this previously collected information was so central to what we produced for this effort, we document it in this chapter. We present this information also to show the variety of wargaming activities covered over the course of our interviews in order to provide the reader with a sense of the diverse wargaming styles and purposes we considered as we created our framework of tasks and sets of recommendations.

Another reason we do this is to provide a more cohesive picture at the center level for how wargames are managed and organized. While the catalog of tools and approaches we previously provided to the Marine Corps contained a considerable amount of information, it was disaggregated to the tool level and so was not able to convey the organizational-level lessons and best practices that were conveyed during our interviews. Stakeholders such as OAD explicitly asked us to gather information about processes and best practices, and this information is best understood within the context of a center and its activities. For the sake of brevity and clarity, the complete write-ups for the wargaming centers are not given here, but can be found in Appendix A, while the catalog of tools and approaches can be found in Appendix B.

This information again largely reflects these organizations' activities in 2016–2017. While much may have changed for individual organizations since then, we believe that this overall snapshot should still be of interest to defense wargamers.

Selection of Wargaming Centers

Working in consultation with the Marine Corps, we put together a list of wargaming centers and other organizations in June 2016 before we began contacting them to discuss tools, approaches, best practices, and other recommendations. This initial list contained 32 organizations in the U.S. defense and intelligence community and in the allied defense wargaming community. This list was not by any means a comprehensive list of the U.S. and allied organizations that conduct wargames.

Several factors went into the selection of this initial list of centers to approach, including sponsor guidance to focus on wargaming centers outside the Marine Corps, as well as a stated interest in naval wargaming. Individual Marine Corps stakeholders also had requests for specific interview locations, including the CNA, the NWC, and SOCOM.

Other factors that went into selecting this initial list of 32 centers were activity level, likelihood of accessibility, and organizational roles. We emphasized wargaming centers that supported enough wargaming activity for there to be lessons learned and other conversations about wargaming tools, approaches, facilities, skill sets, and other aspects. We also identified some organizations on this initial list because the study team had existing professional contacts with them and already knew their wargamers. In some cases, we took advantage of geography, adding centers to our list that were colocated with others we were already planning to visit. We also identified organizations with key roles in wargaming, either formal roles such as through DAWG, or informal roles as established centers that were widely known in the wargaming community.

There are several limitations to this initial list. One is that the list represented our best knowledge of important wargaming activities at the time we put it together. We were also limited by our existing contacts within the wargaming community, which, while fairly extensive, were neither perfect nor comprehensive. This selection process points out a weakness in that we were beginning with centers whose methods and approaches were likely known to us already and who had likely been active in the core wargaming community. This left out centers that were not necessarily active in the wider community, and with whom we had fewer professional ties. This means we ran the risk of merely amplifying what was already general knowledge or conventional wisdom in the established, core community. Another limitation is that wargaming activity over the past few years has been evolving at various DoD organizations. The previous report is unlikely to have fully captured the degree of change that has been happening.

There were also entire sets of wargaming activities excluded by our initial selection. The first area that we did not examine is wargaming within the course of normal staff planning in the military. These are games that military planners would carry out without involving organizations outside their immediate staff. While the centers we did identify carry out games to support operational decisions and plans, this would only constitute a part of the wargaming conducted to support operational planning. The second set of activities that we did not cover include games that most closely resemble live and virtual training simulations. For example, we did not include field exercises, other home station training, or exercises against a live opposing force at locations such as the Army's National Training Center. Our overall sample of wargaming activity in training and education is therefore weighted toward wargaming to support education rather than training. The third area that was only lightly covered by our selection of gaming centers was the area of acquisition wargaming. While a few of the centers we contacted do have games that are considered related to acquisitions, this was not a core focus area in our study. And fourth, we excluded gaming at civilian academic institutions—even those that directly support DoD and other allied government sponsors. This was due to time constraints and the very different nature of organization at civilian universities.

Organization members that we interviewed also identified additional wargaming centers during the course of our discussions. However, due to time constraints, we were not always able to follow up. Some of the notable wargaming organizations we therefore missed include the Air Force Space Command and the Air Staff. We regret the missed opportunity to hear from these organizations.

We eventually contacted or found partial information for all but one of the organizations on our original list. We have center write-ups for 22 of the organizations we contacted while we compiled our catalog of tools and approaches.

The individual center write-ups in Appendix A include the following:

- U.S. Navy
  - NPS
  - NUWC
  - NWC Wargaming Department
  - Office of Naval Intelligence (ONI)
- U.S. Army
  - CAA
  - Center for Strategic Leadership (CSL) at USAWC
  - CGSC
  - TRAC
  - University of Foreign Military and Cultural Studies (UFMCS)
- U.S Air Force
  - AFMC
  - Air Force Research Laboratory (AFRL)
  - Air University LeMay Wargaming Center
- OSD
  - CAPE
- Joint Staff
  - Center for Applied Strategic Learning (CASL) at NDU
  - J-8 SAGD
- CCMDs
  - SOCOM Wargame Center

- Federally funded research and development centers
  - CNA
  - RAND
- Allied Wargaming Centers
  - Center for Operational Research and Analysis (CORA), Defence Research and Development Canada
  - Defence Science and Technology Laboratory (Dstl), UK Ministry of Defence (MOD)
  - Defence Science and Technology (DST), Joint and Operations Analysis Division, Australian Department of Defence.

We also contacted individuals at the Council on Foreign Relations, the Johns Hopkins University Applied Physics Laboratory (JHU-APL), the Modeling and Simulation Coordination Office (M&SCO), the UK Defence Academy, the U.S. Army Training and Doctrine Command (TRADOC) G2, and the intelligence community for information about specific individual tools or approaches.[1] However, we had insufficient information to do center write-ups for these organizations.

There were other organizations that ultimately did not make our list of wargaming centers to contact because they did not have significant wargaming capabilities. We determined through speaking with individuals at both ONA and OSD Policy that while both offices are among the DAWG quad-chairs and act as important wargame sponsors, their wargames were largely contracted out rather than being conducted through an in-house capability. Our conversations with members of the Institute for Defense Analysis and others familiar with it led us to conclude that the institute did not have a significant wargaming capability at the time of our interviews. Neither were we familiar with any wargaming capability at the MITRE Corporation at the time of our interviews.

Format for Individual Center Write-Ups

For each organization we contacted, we briefly describe the organization's mission and how it uses wargaming. We also provide the information we had available on each organization's tools, approaches, processes, skill sets, best practices, and any recommendations. All errors are ours; the situation at any particular organization may have changed in the time since we contacted it, and we no doubt have only partial awareness of any organization's complete set of wargaming activities.

We also use the following general categories of wargames to describe the wargaming at different centers, as stated in Chapter 2:

- wargaming to support concept development
- wargaming to support capability development and analysis

---

[1]   Some members of the intelligence community consider wargaming to be an imaginative thinking technique, one of the categories in U.S. Central Intelligence Agency, *A Tradecraft Primer: Structured Analytic Techniques for Improving Intelligence Analysis* (Washington, D.C.: U.S. Central Intelligence Agency, 2009).

- S&T wargaming
- senior leader engagement and strategic discussion
- wargaming to support operational decisions and plans
- wargaming for training and education.

Major Themes

There were a number of major themes that arose in the center interviews. Centers, including those situated within academic institutions, supported wargaming for a variety of purposes. They also employed a wide range of tools, approaches, and adjudication methods from those of very low cost and low technology to those more technologically advanced, usually depending on the resources available to the wargaming center. Centers used a variety of skill sets to support wargaming, again with better-resourced centers able to support more specialized staff. Consistent with wider trends in the wargaming CoPs to make wargaming practice more explicit, many centers also had or were in the process of writing wargaming guides.

The range of tools discussed by the centers in Appendix A and laid out individually in Appendix B ranged from manual board games, to video software, to DoD program of record (POR) tools, to multiple networks at multiple levels of classification. Centers named both quantitative and qualitative analysis tools, commercially available tools for visualization and collaboration, a variety of structured group methods and adjudication styles, and a variety of reference materials. Even centers that had a fairly high level of resources used a mix of manual and software-heavy approaches. In terms of adopting new tools, centers were continually exploring new commercial software and tools and were further developing government off-the-shelf (GOTS) or custom-built software. On the other hand, there was movement away from POR tools due to issues such as lack of flexibility. Manual games were continually in use regardless of what other technologies centers were exploring and were considered to have significant pragmatic advantages.[2]

In terms of staff and skill sets, the wargaming centers employed a mixture of wargame designers, military operators, OR analysts, software programmers, and information technology (IT) support. Certain organizations such as the NWC and RAND incorporate a higher number of social scientists into wargaming. Many of the wargaming centers stated the growing demand for wargames at higher levels of classification and suggested that it would be invaluable for the Marine Corps to invest both time and funds in acquiring Top Secret Special Access Program (SAP)/Sensitive Compartmentalized Information (SCI) clearances for its wargamers.

---

[2]   The advantages of manual games include cost, flexibility, adaptability, adjudication transparency, and the existence of large libraries of existing manual game mechanics that could serve as references. Manual games were also considered more reliable for classified cases where security and IT policies complicated the use of outside software tools.

The greatest variance between wargaming centers was in the facilities. On one end of the spectrum, some centers did not own any facilities aside from shared conference rooms or classrooms. On the other end, some centers, like the NWC Wargaming Department, possessed a dedicated building, auditorium, configurable rooms, server rooms, and video studios. Nevertheless, all the centers emphasized the importance of configurability in their facilities. As different games require varying requirements in terms of facilities, it is immensely valuable to be able to adapt those facilities in response to each game. The level of configurability may include classification spaces and breakout rooms for team discussions. For wargaming facilities, adaptability and flexibility remain paramount.

Assessing the State of the Art in Wargaming

During our initial discussions about the study, the Marine Corps asked us to also assess the state of the art in wargaming. We understood this not only to mean the technological state of the art, but also trends and progress in methods and approaches. We identified these cross-center and community-wide trends through our interviews and other reference material we collected about center-level wargaming activity, as laid out in this chapter and in Appendix A. We identified items mentioned repeatedly and by multiple centers as trends. We also briefly examined trends in the commercial gaming world, which were analyzed in Chapter 3.

Across our interviews, several characteristics emerged that define the state of the art of wargaming. One is that the adoption of technology in DoD and allied defense wargaming is considerably lower than in the commercial gaming world. We do not predict that defense wargaming, apart from applications that resemble virtual training, will ever be in a position to match the technological investment in current video games, social media games, and other contemporary digital games. This is based on our discussion of the relative size of the digital versus board game markets in Chapter 4, and the information we presented about the scale of defense wargaming. Looking at the available information on the commercial gaming market compared with the scale of activity represented by professional national security gaming, we conclude that the vast user scale available to commercial games, the need for realistic adjudication of complex national security phenomena, the greater need for specialized expertise in professional games, and the degree of customization needed for national security games will cause the latter to remain labor intensive. Classification issues, restrictions on government information systems, and other security issues also pose significant barriers to the use and adoption of commercial game technology. The professional wargaming community is experimenting with AR and VR, and analysis and information management tools with greater capabilities, as discussed in our center write-ups. However, the overall pace of technological change appears modest and at the margins compared with that of the commercial sector.

An important trend in defense sector wargaming that we did notice in multiple interviews is a drive toward systemization: Whether in game design, facilitation, data collection, or analysis, centers are paying more attention to methodical, transparent approaches to gaming in the hope of generating more credible insights. At the same time, there is a concern that games be appropriately tailored to the purpose rather than generating "cookie-cutter" approaches that can be replicated without considerable alteration. There was also an interest in player experience that focused on creating immersive environments where players have meaningful autonomy. Finally, there is increased concern with integrating games into broader efforts in a "cycle of research" in the hopes of improving the impact of games on senior leader decisions, and particularly those relating to plans, programming, and budgets.[3] These characteristics were often aspirational—no one center modeled all of them in all of its work—but they were common concerns that guided the work of many of the most respected gamers we spoke with.

The drive toward systemization is not surprising; it was called for in Deputy Secretary of Defense Robert Work's 2015 memorandum. However, the range and depth of the efforts as evidenced in our interviews and the activities by the wargaming CoPs shows commitment to developing rigorous techniques to support wargaming. In some cases, these focused on the game design processes, including developing solid game objectives, and selecting game mechanics to help achieve these goals. For example, entities such as the NWC, the UK MOD, and USAWC have produced handbooks to help new staff understand the design process and common considerations. Other organizations adopted approaches such as manual board game mechanics and matrix gaming that imposed processes on gameplay and adjudication, making the outcomes of player decisions more transparent and easier to document. We also saw facilitation techniques to structure game discussion that built on work in "soft" OR, red teaming, structure analytic techniques, and strategy development to better draw information from game participants. These tools were seen as particularly helpful when conducting seminar games that might otherwise have had little formal structure. Efforts to draw particular information from players are supported by new approaches to data collection, such as the NPSs data collection and management plan, that devote far more attention to planning out what information needs to be collected as part of game design. Finally, postgame analysis increasingly draws on interdisciplinary tools to assess qualitative and quantitative evidence from the game rather than simply reporting a narrative of game decisions. While these innovations touch on very different elements of the wargaming process, they all represent attempts to structure what information is elicited from players and how it is recorded and analyzed.

At the same time that much energy is being given to structuring aspects of wargaming, there is also a concern that the flexibility of games be preserved to ensure

---

[3] Perla, 2012, p. 252.

that they are tailored to specific objectives. This is standard advice in most wargaming instruction, but also applies to the selection of the spaces and tools that support wargaming. For example, many centers stressed the advantages of gaming spaces with flexible rooms that could be reconfigured to support different game sizes, number of cells, and information flow between rooms. Similarly, computer aids like Standard Wargaming Integration and Facilitation Tools (SWIFT) or the Joint Seminar Wargaming Adjudication Tool 2 (jSWAT2), which could be easily repurposed to games at different levels of analysis, were seen as more helpful then POR models and simulations that were seen as harder to adapt to new epochs, levels, or data requirements. While such spaces and tools were often less flashy (while still being expensive), they saw far more use then more complicated systems that proved to be suitable only for one or two styles of game.

Innovations to support positive player experience were also a hallmark of state-of-the-art games and another trend we saw across interviews. In some cases, this manifested in the use of technology such as video, touch panels, or even three-dimensional projections to provide information in a lively way. However, other centers achieved similar ends by focusing on narrative and storytelling in games using tools like matrix gaming. Regardless of how the effect was achieved, immersive gameplay drew players into the game world, increasing the amount of work players were willing to do. However, this was valuable only if pregame research and player knowledge created a useful and accurate world. Many warned that without systematic approaches to game design, facilitation, adjudication, data collection, and analysis, immersion could create wargames that persuasively produce questionable results.

Finally, members of many centers we spoke with stressed that games had the most impact when embedded in a series of games, a cycle of research, or other broader processes to inform DoD decisionmakers. Series of games are becoming increasingly common, as multiple games allow for different aspects of a problem to be explored and can provide multiple points of comparison. In order to integrate games with other tools, centers stressed that considerable forethought was needed to design multiple games with other types of exercises and analysis so that they build on each other over time. Supporting these efforts requires a diverse team with different types of academic and applied experience. For example, it was common to see game design teams that included current or former military operators or government civilians, social scientists, OR analysts, and a range of programmers, each of whom brought different experience. Such mixed teams not only produced a wider range of games but also were able to advise on how certain games could be best integrated with other games or types of analysis to have greater impact.

Another high-visibility issue for the wargaming community at the time of our interviews was the question of wargaming's impact on programmatic decisions—that is, decisions on what the Pentagon buys. It is important to note that even when games are well integrated, it is very difficult to demonstrate a strong connection between wargames and decisions for the Program Objective Memorandum. Because games are only one aspect of many inputs, tracing the effect of any one tool is extremely difficult. Games are often most effective when they change the understanding of players, something that is not always accomplished through formal briefing and reporting processes. As a result, the effects of games may be quite indirect, and many cautioned against the use of rigid standards or metrics of impact to assess their influence.[4]

4   Alternatively, wargames may affirm the beliefs that players already have about various systems and capabilities. Thus, rather than causing beliefs to change, the games may merely reflect prevailing beliefs and support for certain systems and capabilities that then go on to receive funding.

CHAPTER SIX

The Framework of Wargaming Capabilities

In this chapter we discuss the framework of core tasks common across all the wargame categories, as outlined in Chapter 2, as well as tasks that are specific or particularly central to each wargame category. By breaking down wargames into categories, identifying the tasks associated with the different types of wargames, and identifying ecosystems of tools, we are able to create our recommendations on tools and approaches given the types of wargames that the Marine Corps wishes to emphasize in its future wargaming capability. In this way, we provide a cohesive framework for our major findings and recurring themes, which are later structured in our COA recommendations in Chapter 7.

The framework begins by outlining the core tasks of any wargame, as illustrated in Table 6.1. These are the baseline tasks associated with any wargame, regardless of purpose, time orientation, topic, or adjudication approach. Although debates within the wargaming community often focus on the differences between types of games, there appear to be a common set of activities performed by all wargamers.

We identified these core tasks for all the categories of gaming we examined by drawing from our interviews, published DoD wargaming handbooks, several center briefings about their wargaming processes, and the study team's knowledge of wargaming tasks from our own experience with wargaming. We also shared this framework with the sponsor and certain wargaming centers before drafting our recommendations and report.

Next, in the series of tables that follow, we take these common wargaming tasks from Table 6.1 as a baseline and add or emphasize tasks for each category of wargaming. For brevity, we only list the tasks particularly important to each category of wargame. Complete versions of wargaming tasks can be found in Appendix C. In devising these tasks, we again relied on the information that the different centers had shared with us about their wargaming processes and best practices, as well our own experiences with different types of wargaming. The purpose for providing these frameworks is to make the catalog of tools and approaches in Appendix B more usable by explicitly identifying the wargaming tasks involved so that the reader may have more information to make choices between tools and approaches.

Table 6.1
Core Wargaming Tasks

Task	Description
Understanding sponsor requirements	Understanding the sponsor's objectives for the wargame and analysis.
Managing information	Acquiring and coordinating research among the wargame team.
Understanding the problem	Framing the problem to inform game design.
Developing and managing event process	Coordinating logistical concerns involved with the game such as venue, participants, and classification.
Scenario development	Crafting a scenario designed to inform and draw out the necessary decisions for participants within a game.
Game development	Developing a game through design and scenario validation, rule development, playtesting, and refinement.
Providing facilitation	Facilitating the game to keep players engaged and game events or discussion progressing.
Data capture	Capturing data and insights from the wargame, either through note takers or other tools.
Wargame analysis	Analyzing the wargame's insights, results, and date.

Tasks for Wargames to Support Concept Development

Utilizing Table 6.1 as a baseline, we provide the tasks we believe exist for concept development wargames, but not necessarily for other types of games. In Table 6.2, these additional tasks for concept development wargaming are eliciting or identifying new concepts, developing consistent assumptions about the future, operationalizing the concepts in the wargame, game system design, adjudicating novel concepts, and identifying key uncertainties. We see how the future orientation of many concept development games in particular drives some of these additional tasks, since this category of game often deals with concepts that do not exist in real life. Making future assumptions explicit and translating the concept of interest into the wargame are important. There is also the difficult task of adjudicating the outcomes of employing novel concepts and actions in a fair and transparent manner. Otherwise, players or concept developers may deem the wargame's results and analysis as a poor depiction of the concept's ideas.

Tasks in Support of Wargames to Support Capabilities Development and Analysis

In Table 6.3 we turn to the tasks identified for wargames that support capabilities development and analysis. (Analysis is often understood within DoD to be largely quantitative in nature.)

Table 6.2
Concept Development Wargaming Tasks

Wargaming Task	Description
Eliciting or identifying new concepts	Eliciting or identifying the new concepts to test within the wargame.
Developing consistent assumptions about the future	Developing consistent assumptions about the future, such as the operating environment, technology, capabilities, and political climate.
Operationalizing concepts in the wargame	Creating opportunities in the game for players to use or respond to the identified concepts of interest. This may be done through representing new capabilities, units, different rules, or adjudication adjustments.
Game system design	Creating the processes, rules, mechanics, and visuals for a game.
Adjudicating novel concepts	Adjudicating new or novel concepts that are operationalized in the game but that may not exist in real life.
Identifying key uncertainties	Identifying the key uncertainties in a game—particularly uncertainties about the future and how a new or novel concept may be adjudicated.

Table 6.3
Capabilities Development and Analysis Wargaming Tasks

Additional/Emphasized Task	Description
Retaining the same fidelity as the analytical model	Maintaining the wargame results at a high enough fidelity for data to be compatible with the appropriate analytical model or M&S tool.
Providing data-driven adjudication	Adjudicating the game based on data the DoD analysis community considers acceptable for programmatic decisionmaking.
Generating data for further analysis	Creating data for further analysis to continue the cycle of research.

For wargames that support capabilities development and analysis, we see three additional tasks: retaining the same fidelity in the wargame as an appropriate analytical model, providing data-driven adjudication using data considered acceptable for programmatic decisionmaking, and generating data in the game for further analysis. The back-and-forth between these types of games and further OR-style analyses requires a measure of compatibility between the games and other quantitative tools. This type of wargame is often used to create tactical or operational results that analytical models can further test and examine.

Tasks in Support of Science and Technology Wargaming

Table 6.4 identifies the additional or emphasized tasks associated with S&T wargaming.

Table 6.4
Science and Technology Wargaming Tasks

Additional/Emphasized Task	Description
Eliciting or identifying S&T concepts	Identifying S&T concepts to test within the wargame. Wargames rarely ever generate new concepts, but are useful in experimentation of new concepts.
Developing consistent assumptions about the future	Developing consistent assumptions about the future, such as the operating environment, technology, capabilities, and political climate.
Developing consistent assumptions about the S&T system or concept	When testing a novel S&T system or concept, designers must create a set of assumptions about its capabilities, capacity, and limitations.
Operationalizing concepts in the wargame	Creating opportunities in the game for players to use or respond to the identified concepts of interest. This may be done through representing new capabilities, units, different rules, or adjudication adjustments.
Game system design	Creating the processes, rules, mechanics, and visuals for a game.
Adjudicating novel concepts	Adjudicating new or novel concepts that are operationalized in the game but that may not exist in real life.
Identifying key uncertainties	Identifying the key uncertainties in a game—particularly uncertainties about the future and how a new or novel concept may be adjudicated.

Similar to concept development wargames, S&T wargames involve examining nascent technologies, concepts, and the future operating environment. Thus, many of the core and emphasized tasks are similar to wargaming in support of concept development. However, specific to S&T wargaming, designers must create consistent parameters for experimental or notional S&T systems. This can be achieved through a combination of technical consulting, rapid-turn wargaming as used by the Air Force, and research into analogous technologies.

Tasks in Support of Senior Leader Engagement and Strategic Discussion

We next turn, in Table 6.5, to wargames that primarily aim to engage senior leaders and to generate strategic discussion about an issue. Unlike most wargames, wargaming in support of senior leader engagement and strategic discussion emphasizes facilitated discussion over adjudicating outcomes. As a result, the quality of facilitation and supporting subject matter experts (SMEs) is particularly crucial within this type of game. Moreover, this type of wargame demands a higher level of logistical coordination given the rank of senior leader engagement wargames. Thus, protocol officers can provide significant support in event management. The high-profile nature of these games provides a unique opportunity for advocacy but also presents its own set of challenges.

Table 6.5
Senior Leader Engagement and Strategic Discussion Wargaming Tasks

Additional/Emphasized Task	Description
Identifying key stakeholders	Involving key stakeholders, whether commanders, officials, or SMEs, in the wargaming process.
Accessing protocol officer	Accessing a protocol officer for logistical guidance when working with high-level officers or officials.
Facilitating constructive discussion	Keeping senior leaders in particular on task and participating in constructive discussion.

Tasks for Wargames in Support of Operational Decisions and Plans

Table 6.6 lists the tasks associated for wargames that support operational decisions and plans. Wargaming in support of operational decisions and plans is designed to support the operating forces, such as the commandant commands. Wargaming in support of operational decisions and plans requires an accurate reflection of the operational environment—both in terms of friendly and adversary capabilities and COAs. In particular, COA wargaming is intended to help planners visualize the operational flow, possible COAs, and the operational environment. Joint doctrine on planning emphasizes wargaming all critical events in a proposed COA.[1] Typically, the results of this type of wargame may feed operational exercises and OPLAN revisions. However, it is important to highlight that, under time pressure, the Joint Planning Process often has more in common with red teaming than with formally adjudicated games. Furthermore, this type of gaming requires substantial effort and expertise, often beyond the capacity of operational staff.

Table 6.6
Operational Decisions and Plans Wargaming Tasks

Additional/Emphasized Task	Description
Defining the operational environment	Scoping the operational environment in terms of context, capabilities, the adversary, and other operational factors.
Creating consistent and transparent adjudication process	Creating an adjudication process that is transparent and acceptable to operational planners.
Assessing the real-life OPLAN	Assessing the strengths and weaknesses of the real-life OPLAN within the wargame.
Developing an understanding of friendly and adversarial COAs	Developing and providing a realistic depiction of both friendly and adversarial COAs in terms of doctrine, capabilities, and capacity.

---

[1] Joint Chiefs of Staff, *Joint Planning*, Washington, D.C.: Joint Publication JP 5-0, June 16, 2017, p. V-31.

There are three tasks that we emphasize in particular for wargames that support operational decisions and plans: defining the operational environment, creating a consistent and transparent adjudication process, and developing an understanding of friendly and adversarial COAs. While other types of games require understanding of the operational environment, transparent adjudication, and an understanding of COAs, these steps are particularly important when supporting operational decisionmaking. A fourth task, assessing the real-life OPLAN, is unique to this category of gaming.

Tasks for Wargames That Support Training and Education

Table 6.7 lists the tasks required for training and education wargames, highlighting the additional tasks pertinent to this category.

Wargames in support of training and education differ from other games because they have specific learning objectives or classroom curricula with which they should align. Certain additional tasks for this category, such as providing an immersive experience, providing feedback to students, and evaluating learning, arise from the classroom environment.

The Ecosystem of Tools

In the course of our interviews with members of wargaming centers, we included both manual and software-based tools within our catalog and our analysis. However, there is particular interest within parts of the Marine Corps on software-based tools to support wargaming, so we give some additional attention to those here. For the purposes of this report, we generally categorize the software tools in particular into different ecosystems with some commonalities in support requirements, developer type, and

Table 6.7
Training and Education Wargaming Tasks

Additional/Emphasized Task	Description
Understanding learning objectives and curriculum	Understanding the learning objectives and curriculum of the class and aligning the game with them.
Drafting or co-opting a game for learning objectives	Creating or adapting a game around learning objectives.
Providing an immersive experience	Providing an immersive experience in which students must assume the roles they play within a game. Gameplay should be engaging and fun.
Providing feedback to students	Providing feedback to students on their gameplay.
Evaluating learning	Evaluating the level of student learning from a game in relation to the learning objectives and the class curriculum.

a few other characteristics. We identify the four general tool ecosystems as commercial off-the-shelf (COTS), GOTS, PORs, and custom-built. COTS is self-explanatory, while POR tools are those funded by specific line items in the defense budget. GOTS and POR tools are both funded by DoD, but the former often lack a budget line item. GOTS tools also tend to be smaller and less widely used than POR tools, though that is not always the case. There is also overlap between GOTS tools and those that a government organization may have developed in-house for its particular needs. In these cases we consider a tool to be custom-built rather than GOTS if it has not previously been made available to other government or industry users. We make this distinction because there is likely some history of transferring tools to other users with GOTS, but not necessarily with custom-built ones.

Each ecosystem and its underlying approach to acquiring and developing tools possesses its own advantages and disadvantages. For example, there are several advantages to adopting COTS tools. One obvious advantage is that they offer instant capabilities without development risk. Trial versions also allow potential customers to try out software before making purchase decisions. COTS tools also tend to maintain large user communities, with extensive customer support, compared with government-originated software. To maintain their user communities, commercial developers also frequently update their COTS tools over successive years. However, COTS tools generally offer less customization and do not always focus on military applications to the same extent that DoD-initiated software will. More important, due to government IT policies, COTS tools can be restricted from government systems—especially classified ones. COTS software also typically requires a review process before it is approved for use on government systems. There are also open-source tools with large user communities and organizations dedicated to improving the tool.

Relying on POR tools is another approach. POR tools often possess existing data and users within the government. This allows for a distinct advantage of familiarity and accessibility for government organizations in their use. However, POR tools can require substantial investment in terms of facilities and staff. Heavy M&S tools can also require significant time and manpower to develop the scenarios and data required to run them. Wargamers have also noted the greater usefulness of some POR tools for training rather than wargaming, noting that such tools can be less flexible for wargaming future scenarios.

Adopting GOTS and other government-developed tools with a user base outside that of the initial developer is another approach. These tend to be more specialized for defense wargaming purposes than COTS tools, but many times they come with fewer facilities and less staff commitment than some POR tools. There are a fairly wide number of GOTS tools and approaches that we encountered during the course of our discussions with the wargaming community. Examples include the Joint Wargame Analysis Model, or JWAM (formerly known as the Center for Army Analysis Wargame Analysis Model); SWIFT; and the Versatile Assessment Simulation Tool (VAST). The advantages here are demonstrated applicability to military wargaming, development that has already been

Table 6.8
Ecosystems of Wargaming Tools

Primary Ecosystem	Advantages	Disadvantages
COTS	• Larger user community • Customer support available	• Not all are allowed on government networks • Less customization • Less developed data or scenarios for military
POR	• Well understood • Often existing data • Existing users	• Reduced flexibility • Less used in wargaming • Some require large footprint (facilities/people)
GOTS	• Already developed • Free to government users • Existing users	• Requires knowledge of tool • Sometimes limited user community
Custom-built	• High degree of customization possible	• Requires time and money to develop • Redundancy with existing tools

funded by the government, and often a user community of other government wargamers. Disadvantages of adopting GOTS rather than COTS tools include a smaller user base than for popular commercial tools and limited or no formal available training for newcomers to the tools. Access to other government users therefore becomes more important.

Finally, another possible approach is a custom-built ecosystem that allows for the highest level of customization and flexibility. The NWC web applications and LeMay Center Wargaming Gateway are examples of custom-built ecosystems. The developer can tailor tools to the organization's specific needs. At the same time, the custom-built ecosystem requires a substantial investment in resources and time, and we do not recommend that the Marine Corps custom build its wargaming tools. Custom-built tools can also run the risk of duplicating the functionality of commercial- or government-developed tools that were already available, thus saving additional development time. Table 6.8 summarizes the major tool ecosystems we have described.

Any tool ecosystem or combination of ecosystems requires resources for setup, integration, and maintenance. Regardless of the tools, additional resources will be necessary to adapt and customize new tools for Marine Corps scenarios, data, and concepts. The wargaming centers at which we conducted interviews often used a combination of ecosystems, with larger centers such as the NWC using a wide variety of COTS, custom-built, and several POR tools. Thus, the ecosystems of tools are designed to provide a coherent method of approaching and building a software-based wargaming capability for the Marine Corps. This does not mean that the following COA recommendations in Chapter 7 exclude manual tools and techniques. On the contrary, the following recommendations include a myriad of manual tools, methods, and approaches serving as a critical foundation for the Marine Corps' wargaming capability. The ecosystems of tools serve as key decision points for the Marine Corps as they pursue their NGW concept and future wargaming capabilities.

CHAPTER SEVEN

Courses of Action

In this chapter we discuss the COAs we recommend for Marine Corps investment in its future wargaming capability. These are based on our previous review of wargaming tools and approaches in use; visits and interviews at other wargaming centers; our understanding of Marine Corps priorities in wargaming; the associated wargaming tasks that we anticipate the Marine Corps will have to handle; and other considerations such as fiscal cost, risk, time, and difficulty of implementation. We begin the chapter with implications for Marine Corps wargaming in relation to the state of the art of wargaming as assessed in Chapter 5, our interviews at wargaming centers, and broader gaming trends as discussed in Chapters 3 and 4. Then we explain how we approached developing the COAs and the subsequent three-tier COAs, demarcated by the level of resources—low, medium, and high. The chapter concludes with key takeaways and next steps in building the Marine Corps' NGW concept and wargaming capability.

Implications for Marine Corps Wargaming

Considering our interviews, the state of the art in wargaming, and broader gaming trends, there are several implications for Marine Corps wargaming. One of the most significant implications for Marine Corps gaming is the fact that the state of art is a mixture of various processes, methods, and tools and is not defined by a particular set of technologies or a set system. Thus, the Marine Corps should strive to adopt, adapt, and evolve the current best practices in the wargaming community in its pursuit of NGW. This inevitably involves the progressive reformation of its entire wargaming capability from processes, personnel, methods, tools, and facilities. There is no way for the Marine Corps to outright purchase its vision of NGW. Realizing its vision will require experimentation, concerted effort, patience, and long-term vision on the part of the Marine Corps.

The Marine Corps must also take into account the potential commercial sector developments and generational experiences with gaming that may affect how it should wargame in the future. Given that some rudimentary elements of NGW do already exist outside national security wargaming, particularly in the commercial gaming world and in civilian academic programs, the Marine Corps can begin to experiment and adopt numerous processes and methods already in use by the wider gaming community. This can involve examining other wargaming centers' manuals and standard operating procedures, experimenting with design thinking, or executing trials on various models like JWAM. Fortunately, this comes at a time when the wargaming community is working toward formalizing wargaming education and the practices it has used to date.

Internally, the Marine Corps can begin experimenting with no-turn games and other potential future directions in addition to adopting known best practices from organizations already experimenting with no-turn gaming constructs. We identify three primary elements of a no-turn game that the Marine Corps should focus on: (1) simultaneous player actions, (2) real-time gameplay, and (3) disaggregated adjudication. It is also important that the outcome of actions is determined in a consistent way to ensure that players get appropriate feedback.

Next, given the broad generational constructs of gaming, the Marine Corps should explore options through which it can establish a system of educating and hiring young, junior wargamers. Fresh, new perspectives from younger generations could prove invaluable in enriching lessons learned by previous generations of gamers. This may include new methods and ideas on gameplay, technology integration into gaming, and game design. Additionally, the Marine Corps' concept of NGW will inevitably require a new generation of gamers to adopt and execute its vision of wargaming in the future. By establishing a method of cultivating wargaming education and a new generation of wargamers, the Marine Corps will be better positioned for enduring future success.

Finally, we warn against the impulse for the Marine Corps to imitate commercial gaming, and particularly commercial video games. While, admittedly, military wargames have largely remained stagnant in their manual, board game–style format, commercial games have made tremendous technological leaps. As an industry, the commercial gaming sector, and particularly digital gaming, has embraced a style of fast-paced, visually immersive gaming, complemented by real-time adjudication and gameplay. On the surface, military-style games like *Harpoon* (1989–2007), *Age of Empires* (1997–2013), *Hearts of Iron* (2002–2016), *War in the Pacific* (2004), *Rome: Total War* (2004–2013), and *AirLand Battle* (2013) embody the type of immersive, elaborate gaming the NGW envisions. Therefore, it is understandable that military wargamers may be tempted to imitate their commercial counterparts, especially as younger generations of gamers come of age.

However, translating the commercial style of gaming and its associated technologies is a difficult endeavor with potential pitfalls. First, despite their veneer of warfare, the large majority of military-style commercial games are poor imitations of actual warfare, lacking the rigor and historical accuracy required for professional wargaming. Commercial games often ignore key considerations in warfare like terrain, command hierarchy, the use of formations and fortifications, combined arms warfare, and logistics. For instance, *Age of Empires* lacks historical fidelity and any semblance of strategic thinking; similarly, *World of Tanks* (2010–2014) is a glorified first-person shooter game involving tanks. Overall, most commercial military-style games are fundamentally war-themed, which are simplified to a level of abstraction that is appealing to mass-market audiences.

Admittedly, some games do achieve a greater sense of realism and strategy. A good example is *Rome: Total War*. Gamers play as a specific faction in the ancient world, ranging from a prominent Roman family to Egyptians, a common gaming trope. However, unlike most games, *Rome: Total War* incorporates intricate factors like troop morale, terrain, specific unit formations, taxation, and even diplomacy. Gamers must constantly switch from strategic concerns like tax policy and alliances to tactical battles where a player commands thousands of animated soldiers, who move according to flocking algorithms. For modern warfare, *Hearts of Iron III* (2009), a World War II–based game, features comparable intricate gameplay strategically and operationally. For instance, humidity affects the effectiveness of air operations, logistical concerns can cripple military campaigns, and units possess specific strengths and weaknesses—all of which is facilitated by artificial intelligence and complex computer algorithms. Even so, games like *Rome: Total War* and *Iron Heart III* are the exceptions, not the rule.

This is not to say that commercial gaming firms could not create realistic wargames. The industry possesses the technological and gaming expertise to craft a truly stunning and realistic wargame (albeit with the advice of military SMEs). However, unlike professional wargamers, commercial firms are motivated by unit sales and popularity, not accuracy and analytical rigor. Commercial games, whether war-themed or not, often sacrifice accuracy and realism for playability—resulting in a common denominator approach to game design. Therefore, if the military truly wants to leverage the commercial sector in wargaming, professional wargaming will have to co-opt premier commercial firms into defense work, build its own technologically advanced capability, or promote independent developers to provide niche games. At any rate, all of these options will require a substantial, if not prohibitive, level of funding for the small defense wargaming community. As the Marine Corps attempts to build its NGW capability, the service must be wary of confusing technological advances with inherent gaming advances. Explicit recommendations and COAs are further explained in the following sections.

Developing Courses of Action

The WGD requested low-, medium-, and high-resourced COAs in terms of time, financial commitment, and difficulty of implementation. In devising these COAs we identified current best practices in the wargaming community, then compared them with Marine Corps priorities in wargaming and its vision for NGW. In other words, we made recommendations based on practices that had already been shown to be successful at other organizations or were explicit recommendations made by multiple centers to pass along to the Marine Corps. The COAs are a combination of state-of-the-art techniques, methods, and tools in wargaming that best facilitate the implementation of NGW. We constructed the COAs to build on one another in a progressive manner, demarcated by tiers of fiscal cost, risk, time, and complexity of execution.

The low-resourced COA encompasses the actions that the Marine Corps could take with minimal additional resources, focusing mostly on reforming and adopting processes. The medium-resourced COA, which requires additional resources, mostly looks at acquiring additional skill sets and equipment. Finally, the high-resourced COA adds the actions that could be taken if a very high level of resources were available, mostly looking at constructing additional wargaming-specific facilities. Guidance from our sponsor was that detailed cost estimates were outside the scope of this effort. We therefore make some approximations about what would be more resource intensive and assume that building new facilities is the costliest action of the wargaming-related options available to the Marine Corps.

In our process of developing these COAs, we sought to mitigate risk for the Marine Corps by putting additional emphasis on tools, approaches, and practices that had proven to be useful to other wargaming centers. At the same time, we attempted to make recommendations that would still support the development of no-turn gaming and other elements of NGW. Further development of staff and providing additional research and development time are likely to be key to getting no-turn gaming started. Again, we have focused on wargames to support concept development, capabilities development and analysis, and senior leadership engagement and strategic discussions. We also make some broad recommendations about types of investments rather than specific, individual tools. We also assume that the future wargaming capability will develop over time.

It is important to stress that the COAs are cumulative, reflecting a progressive methodology. Building facilities without a foundation in processes, methods, and tools will not allow the Marine Corps to reap the full benefits of a new wargaming center. The COAs are illustrated in Figure 7.1. We recommend that the Marine Corps implement as many COAs as it has resources for, beginning with the low-resourced COA. If there are few resources available for improving wargaming, we recommend the low-resourced COA. If there are resources enough to support expenditures such as constructing facilities, we recommend that the Marine Corps implement all the COAs. Again, it is vitally important to stress that the COAs build on each other—they are not alternative choices. Because of this, we do not include advantages and disadvantages as we would for discretely different paths.

Figure 7.1
Low-, Medium-, and High-Resourced Courses of Action for Marine Corps Wargaming

High-resourced COA
- Builds upon the two previous COAs, but incorporates new facilities

Medium-resourced COA
- Builds upon low-resourced COA, but incorporates new staff to support more intensive tools and methods

Low-resourced COA
- Involves changing processes and methods without improvements to staff or facilities

Courses of Action

The Low-Resourced Course of Action

The low-resourced COA emphasizes changes to processes and methods without increasing staff size or building additional facilities and reflects the best practices of many wargaming centers. This COA also tries to maximize wargaming capabilities with minimal additional resources by focusing on "low-hanging fruit."

Our recommendations for the low-resourced COA are as follows:

1. Continue Marine Corps–level integration of wargaming with capability development and programmatic decisionmaking. This integration should increase the inclusion of wargaming into the Marine Corps capability development and analysis process, and will have a greater impact than the adoption of any particular tool or approach within individual wargames.

2. Develop rapid- or quick-turn games that can explore issues in a responsive way by leveraging small groups and flexible game designs. This process can be supported by contracting support by labor time rather than a set number of games to provide operational flexibility to the center. This was suggested during the course of some of our discussions at other wargaming centers. Current contracting requirements at the WGD means that support is provided to a certain number of wargames per year.

3. Bring in additional low-cost resources to support innovative gaming. Based on practices at other wargaming centers, these might include spreadsheet tools, materials to create board games and to modify commercial board games, and a collection or library of commercial games and wargaming books.

4. Begin a GOTS ecosystem through the adoption of one of the many GOTS wargame support tools available. We recommend experimenting with different tools and adapting and further developing them for Marine Corps purposes. GOTS tools include software such as SWIFT or VAST, processes such

as JWAM, government board games such as Synthetic Staff Ride (SSR): Mind-anao, and wargame development aids such as the Caffrey Triangle, which serves as a framework for considering the purpose of a red team in a game. There are also allied tools such as jSWAT2 and the Rapid Campaign Analysis Toolset (RCAT). Adopting GOTS tools that are already in use is also in keeping with the low-resource emphasis of this COA.

5. Provide formal wargaming education to staff on different wargame styles and adjudication methods. There are a variety of resources available, such as AFMC or NPS wargaming classes, MORS wargaming course offerings, UFMCS red teaming classes, and formal facilitation training.

6. Share information and collaborate with other wargaming organizations. Actively participate in other organizations' wargames, either through formal staff exchanges or through informal inclusion in wargames conducted by others.

7. Expand the network of experts to draw into wargames, including faculty at MCU and SMEs available at universities, research centers, think tanks, and other organizations in the greater Washington, D.C., metropolitan area.

The Medium-Resourced Course of Action

The medium-resourced COA represents the set of actions we recommend for building on the actions in the low-resourced COA. The actions here relate more to staff, equipment, and tools.

Our recommendations for the medium-resourced COA are as follows:

1. Bring in additional staff with skill sets in software development or programming; IT and network support; OR, social science, or other analytical backgrounds; and wargame design experience with other defense organizations.

2. Add staff and equipment for multiple networks, including a stand-alone network that is only for WGD use and not connected to other networks. The other networks may include the Non-Classified Internet Protocol Router Network (NIPRNet), the Secret Internet Protocol Router Network (SIPRNet), and the Joint Worldwide Intelligence Communications System. With separate networks come separate servers and machines. We make this recommendation because these networks are necessary to support games at higher levels of classification.

3. Provide staff time to further develop existing tools and to research and prototype with new gaming constructs.

4. Provide Top Secret SCI clearances for staff in order to support events at this higher level of classification.

5. Add COTS tools for knowledge management, data capture, analysis, video creation, and graphic design.

6. Buy a large-format printer, a tool that many wargaming centers find useful in creating maps and visuals for their wargames.

Table 7.1 lists several COTS and open-source tools used by other wargaming centers.

Overall, the medium-resourced COA stresses the addition of staff with specialized skills, staff-led research into wargames, and, in some cases, higher security clearances. This emphasis on staff skills should also come with additional time for staff to

Table 7.1
Examples of Commercial Off-the-Shelf and Open-Source Tools for the Medium-Resourced Course of Action

COTS/Open-Source Tool	Use	Description
Adobe Creative Suite/Creative Cloud	Video/graphic design	A software suite of graphic design, video editing, and web development tools.
i2 Analyst's Notebook	Analysis	Allows users to rapidly piece together disparate data into a single cohesive picture; identify key people events, connections and patterns, increase understanding of the structure and hierarchy of data; and generate visualizations.
ATLAS.ti	Analysis	Toolset to manage, extract, compare, explore, and reassemble meaningful pieces from large amounts of data in flexible yet systematic ways.
Decision Lens	Analysis	An end-to-end software solution and process for identifying, prioritizing, analyzing, and measuring which investments, projects, or resources will deliver the highest returns; and allows organizations to immediately see the impact and trade-offs of the choices they make.
Google Drive and Google Sites	Knowledge management	Creates websites that act as a secure place to store, organize, share, and access information.
Jabber	Data capture	A Windows application that integrates instant messaging, video, voice, voice messaging, screen sharing, and conferencing capabilities securely into one client on a computer desktop.
R	Analysis	A programming language used for statistical data analysis and visualization.
Statistical Package for the Social Sciences (SPSS)	Analysis	A predictive analytics software tool that allows analysis; data management (case selection, file reshaping, creating derived data); and data documentation.
Systems Tool Kit	Analysis	A modeling environment used to display and manipulate air and space systems.
Tableau	Analysis	A software tool that allows the visualization of large and complex data sets and multidimensional relational databases.
ThinkTank	Knowledge management	A group decision support software for brainstorming, innovation, decisionmaking, and virtual interactive meetings.
Zing Portable Team Meeting System	Knowledge management	Software and hardware package designed for real-time collaboration, where users can anonymously project comments on a screen so they are not spoken over.

learn and experiment with new ideas and perspectives in order to spur innovation in their wargaming. The adaptation of additional COTS tools also depends on having a staff with the specialized knowledge to use them. This is particularly the case with the COTS tools available for analysis.

The medium-resourced COA also recommends equipment: a stand-alone network, enterprise/classified networks, and a large-format printer. Again, because of the restrictions on government networks, a stand-alone network is valuable for experimenting with tools. We note, however, that more networks require more staff to maintain such a network. Centers such as the NWC run as many as seven networks at different levels of classification.

In Table 7.2 we include the two tool ecosystems recommended thus far. The first recommendation is for the Marine Corps to incorporate GOTS tools developed by other government users into its wargames and further develop them for its own needs. Then, as additional resources are available, the Marine Corps should consider the plethora of COTS (and open-source) tools that are available for various needs. We again recommend testing out different tools, seeing what is suited to Marine Corps needs, and incorporating them into the wargaming process over time.

The High-Resourced Course of Action

The high-resourced COA makes further recommendations for facilities and additional network considerations. This is the most resource-intensive COA, one that presumes the Marine Corps has the funds to invest in facilities on top of implementing significant changes in processes, methods, tools, and personnel. We again stress that wargaming processes and the skill sets of the people involved in wargaming are key. That said, there are a few important lessons about wargaming facilities that we were able to gather from other wargaming centers.

Our recommendations for the high-resourced COA are as follows:

1. Consider the highest level of classification that the Marine Corps desires to handle in its wargames. The NWC suggested building to the Top Secret SAP/SCI classification level due to what it observed as an increasing demand for

Table 7.2
Tools Ecosystem for the Medium-Resourced Course of Action

Primary Ecosystem	Advantages	Disadvantages
GOTS	• Already developed • Free to government users • Existing users	• Require knowledge of tool • Sometimes limited user community
COTS	• Larger user community • Customer support available	• Not all are allowed on government networks • Less customization • Less-developed data or scenarios for military

events at that level. However, due to expense, and the need to still have spaces for visitors with no clearances, it may make sense to have only part of any new wargaming facility cleared at that level.

2. Build configurable spaces. One of the key recommendations from multiple wargaming centers is to avoid spaces that are built for a single purpose, such as only having an auditorium with fixed seating. Instead the recommendation is to build spaces with movable and reconfigurable seating, tables, and walls. The aim here is to allow for flexibility and breakout groups.

3. Build configurable networks and audiovisual (AV) equipment that can support the new configurable spaces. The CNA, for example, stressed the need for configurable AV equipment. The NWC also showed us its network cables, which are under the raised flooring in its configurable wargame spaces.

4. Create workspaces for wargame teams. Although there is a (justified) focus on spaces for wargame participants, we also recommend spaces for the staff who design and prepare for wargames. Over the course of our interviews, we saw how the full wargame process often takes months, and workspaces for the teams working on the different games is thus recommended.

5. Furnish a reference library. Our final recommendation for the high-resourced COA is a physical space for the library of commercial board games, PC games, modified games, books, game construction material, and other reference material for wargames.

Takeaways and Next Steps

The Marine Corps' vision for NGW is an ambitious endeavor, seeking to push the boundaries of wargaming through the amalgamation of technology and best practices. Yet attempting to reform an established capability in a revolutionary manner is rarely without challenges. Several obstacles exist for the Marine Corps in executing its concept of NGW. Thus, here we outline some of the potential challenges and possible steps forward.[1]

1. **Under the Budget Control Act of 2011, the DoD budget remains under restrictive guidelines.** Moreover, the utilization of continuing resolutions as short-term budgets for DoD has undercut long-term planning, readiness, and force structure. The prospects of a long-term budget remain contentious and unlikely at the time this report is being written. Partisan divisions pose significant obstacles toward a comprehensive budget for DoD. This is compounded by other ambitious acquisition projects across DoD, such as the expansion of the naval fleet.

---

[1] This reflects the situation with Marine Corps wargaming in 2017.

2. **The Marine Corps must strive to further integrate its wargaming capability into the operating forces.** Currently, wargaming helps develop capabilities and capacities, but it is not intimately intertwined with Marine Corps operating forces in their training or planning processes. Whether through an operational red teaming relationship like the one at SOCOM, the Marine Corps should actively attempt to socialize wargaming and its utility throughout the service. The staffs at the various Marine Expeditionary Forces headquarters should be active and regular participants in the WGD's wargames, and the WGD should actively solicit challenges and suggestions from the Marine Expeditionary Forces.

3. **Provide a long-term execution plan for NGW.** Many ambitious projects and initiatives have stalled or disappeared within a single administration. Under Marine Corps Commandant James F. Amos, red teaming was considered a special effort for the service. However, after his tenure, the red teaming initiative largely dissipated. Therefore, it is best to seek concrete steps in the plan of actions and milestones process for NGW for long-term success.

4. **Navigate IT policies and facilities regulations.** Any technological acquisitions and upgrades will require compliance with a myriad of IT policies, including a directive to eliminate stand-alone networks. The challenge will lie in balancing flexibility of thought and execution in wargaming with maintaining viable cybersecurity and compliance.

5. **Expand and nurture institutional knowledge in wargaming.** In collaboration with MCU, the WGD should expand its educational mission to emulate programs like the NPS's wargaming courses. As such, the Marine Corps can educate the next generation of officers and noncommissioned officers in wargaming and its utility. Ideally, over time, individual marines could eventually obtain secondary military occupational specialties or certifications in wargaming.

6. **Conduct a series of NGW-themed wargames across the DoD.** The notion of NGW will be received with both praise and criticism. Therefore, to best position NGW for success, the Marine Corps should strive to establish buy-in from key stakeholders, policymakers, and other wargaming centers through a series of DoD-wide and joint wargames. This does not simply mean inviting a joint audience. To truly build a sense of ownership of the NGW across the DoD, the Marine Corps should involve designers, action officers, analysts, and participants from across the DoD wargaming community. This will allow the NGW to be socialized, and also to benefit directly from the collective expertise of the wargaming community.

Despite several challenges, there are several current factors in favor of efforts to further develop Marine Corps wargaming. Senior leadership remains enamored of wargaming. The wargaming community is in the process of establishing best practices, exchanging ideas, and actively seeking means to improve itself.

Wargame Center Write-Ups

This appendix contains write-ups for 21 of the wargaming centers at which we conducted interviews during the course of our study. The information presented here represents interviews and visits conducted in 2016–2017. We also shared these write-ups with the different centers in 2017–2018 to check accuracy and get some updates. Again, information on activities and tools for individual centers may have changed between the time of our last contact and the date of this report. However, we believe it is still important to offer this information to others interested in wargaming in order to gain organizational-level perspectives. The intent of this appendix is to impart to those interested in developing wargaming capabilities a sense of a given center's scope, scale, approaches, and tools for gaming in light of its organizational purpose and mission.

We believe that the material presented in this appendix will also be of interest to the defense wargaming community overall as a means to document some aspect of the activities, tools, and approaches of this time period. As interest in defense wargaming waxes and wanes over the decades, we propose that it is useful to document what happens during times of focused interest in order for the defense community to have a better sense of what was and was not typical wargaming practice in previous periods. Many wargaming centers also offered advice on best practices that are generally useful beyond the specific purpose of this study, and we also wanted to document those here.

U.S. Department of Defense Gaming Centers

U.S. Navy Wargaming Centers

We were able visit the NPS, NUWC, and the NWC and discuss wargaming with staff at these locations. We were also able to have a telephone interview with a staff member at ONI about its wargaming activities. The Navy contains many organizations that wargame, and the organizations we contacted are but a few of them. As a reminder, although the Marine Corps falls under the Department of the Navy, we were asked to focus on wargaming outside the Marine Corps. Therefore, there are no write-ups for Marine Corps organizations who regularly wargame.

The Naval Postgraduate School

Overview

The NPS serves as a research university providing unique advanced graduate education for naval and defense personnel and international partners.[1] Additionally, the NPS faculty and students conduct sponsored research in direct support of the Office of the Chief of Naval Operations, the Secretary of the Navy, and various programs throughout the fleet.

Wargaming at the Naval Postgraduate School

Over the last five years the NPS has conducted over 50 wargames, ranging from basic educational games to more advanced analytical games in direct support of DoD. Organizationally, the NPS performs wargames for various sponsors like the Department of the Navy, major commands, and international partners and allies.[2] Wargaming at the NPS spans the field to support concept development, operational decisions and plans, and training and education.

The NPS is unique in our sample of centers in that its primary focus includes teaching wargaming to students rather than simply on conducting wargames as either analytic or educational tools. Because the NPS is an educational institution, the wargaming program strives to combine educating its students in wargaming while adding operational value to the fleet regarding relevant problems. As such, sponsors provide objectives and issues for the NPS to wargame; then NPS faculty and students design, develop, execute, and analyze the wargame.

NPS faculty focus on wargaming as an analytical tool to better understand and explore the implications of human decisionmaking in warfare. Generally, NPS wargaming has three main purposes: analysis, education, and experimentation. More specifically, NPS wargames can be used to develop and assess CONOPS for new platforms and technologies, validate and update existing war plans and tactics, and plan future operations and create new plans. Additionally, the NPS provides technical red teaming or assesses vulnerabilities in emerging blue team capabilities or concepts.

Tools and Approaches

The NPS utilizes a wide range of tools and approaches in its wargames. These include maps, charts, computer visualization, surveys, and computer combat models. However, the NPS as an institution stresses the importance of decisionmaking in wargaming over any specific tool or approach.

Depending on the specific student group, sponsor, and wargame topic, the NPS has used a wide range of techniques and tools to support its wargames. A summary of major tools and approaches are shown in Table A.1.

---

[1] NPS, "NPS Vision," webpage, undated.

[2] Interview with NPS faculty, Monterey, Calif., August 15, 2016.

Table A.1
Naval Postgraduate School Wargaming Tools and Approaches

Tool or Approach	Description	Usage
Joint Seminar Wargaming Adjudication Tool (jSWAT)	Computer-based tool for a seminar wargaming environment, which includes a planning environment to aid in creating synchronization matrices and a simulation used to adjudicate maneuver, logistics, combat, and intelligence gathering.	NPS has fielded jSWAT in a lab based on a memorandum of understanding.
Map Aware Non-Uniform Automata (MANA)	MANA is an agent-based model that allows the exploration of a wide variety of issues with minimal setup time.	This agent-based M&S tool can be useful for exploratory analysis during the early stages of advanced concept development.
Massive Multiplayer Online Wargame Leveraging the Internet	Online brainstorming platform for running discussions with strict format. Players are able to play 140-character "idea cards," which other players can then respond to. Platform includes a blog and other means of pushing wargame updates to participants.	Used for Navy wargaming support.
Modified commercial board games	Use of commercially available games for educational or analytical use. For all but the simplest games, modification is generally needed to make rule sets more accessible and create a playable game to address the educational requirement.	Used for training and education.
NPS Analytic Wargaming course	An 11-week course offered by the NPS that teaches the basics skills required to initiate, design, develop, execute, and analyze a wargame.	Used for education.

Facilities

The NPS has a range of connected classrooms, computer labs, and gaming facilities that can be used for wargaming. Wargames can be conducted at the Top Secret level but are typically limited to the Secret level because of student clearances.

The NPS's ability to teach gaming at a range of levels is one of its critical skill sets. NPS faculty with years of teaching experience offer both a weeklong Basic Analytic Wargaming Mobile Training Team course and an 11-week course on wargaming. These courses provide an opportunity to learn and share institutional knowledge across joint, international, and industry partners.[3] Both courses are built around designing a wargame for a specific sponsor; OAD has been a past sponsor for NPS student wargames.

---

[3] Jeff Appleget and Rob Burks, "Naval Postgraduate School Mobile Training Team (MTT) Wargaming Program," briefing slides, NPS, Monterey, Calif., August 2016; interview with NPS faculty, Monterey, Calif., August 15, 2016.

Figure A.1
Naval Postgraduate School Basic Analytic Wargaming, Major Tasks

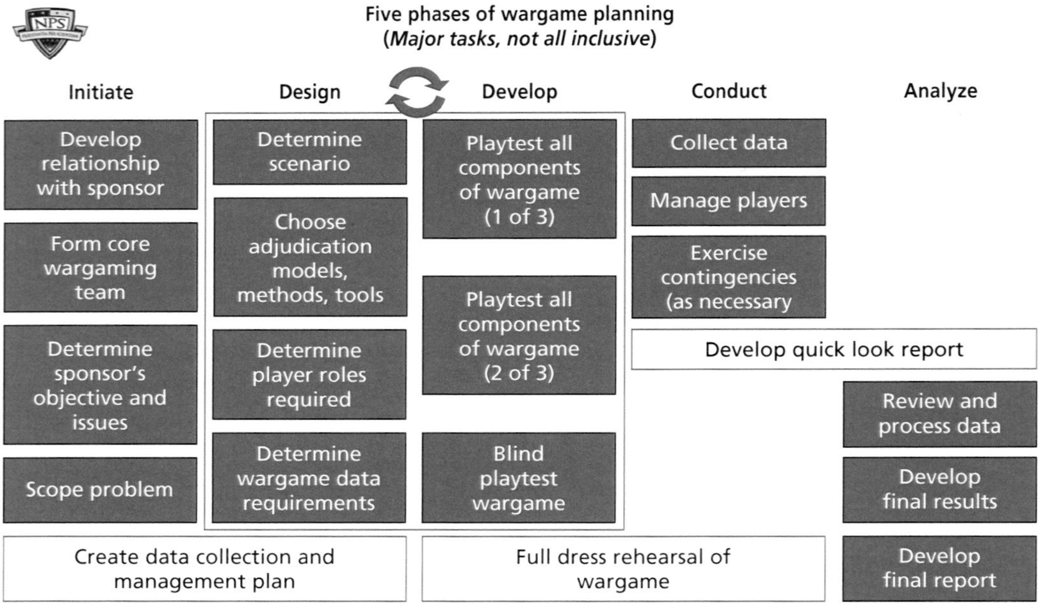

Initiate	Design	Develop	Conduct	Analyze

Five phases of wargame planning
(*Major tasks, not all inclusive*)

Initiate	Design	Develop	Conduct	Analyze
Develop relationship with sponsor	Determine scenario	Playtest all components of wargame (1 of 3)	Collect data	
Form core wargaming team	Choose adjudication models, methods, tools		Manage players	
Determine sponsor's objective and issues	Determine player roles required	Playtest all components of wargame (2 of 3)	Exercise contingencies (as necessary	
			Develop quick look report	
				Review and process data
Scope problem	Determine wargame data requirements	Blind playtest wargame		Develop final results
Create data collection and management plan		Full dress rehearsal of wargame		Develop final report

SOURCE: NPS, Wargaming Activity Hub.

For its wargames, the NPS teaches and uses a process to define objectives with sponsors; create a data collection and management plan; design and develop game scenarios, rules, and tools; and execute and analyze a wargame. Figure A.1 shows a basic outline of the methodology taught by the NPS wargaming course.

Several sessions are spent on the basics of wargame design, including an overview of the design process and examples of different gaming styles and tools. As one of the steps, students devise a data collection and management plan that will guide analysis, develop the "measurement space" of the scenario, data, and methods; examine models and tools that drive the wargame; and define the required rules and players. Of these, the faculty focus is on the data collection and management plan, which deconstructs the sponsor objective into the information that must be collected during the wargame to answer the sponsor's core question.[4]

Unsurprisingly, given the NPS's strong OR curriculum, the school has also devoted considerable effort to the use of OR tools in conjunction with wargames for analysis. Teaching materials devote considerable time to differentiating wargames from M&S and providing examples of how the two approaches can complement each other.

---

[4] Jeff Appleget, "Module 1" and "Module 2," annotated teaching slides, NPS, Monterey, Calif., August 2016; interview with NPS faculty, Monterey, Calif., August 15, 2016.

In particular, the NPS focuses on analytic wargames, where wargame outputs either produce an analysis report for a sponsor or provide input to an M&S effort. For example, COA wargames can provide a concept of operation that is then instantiated into a combat computer simulation to produce a quantitative look at potential outcomes.[5]

Key Best Practices and Recommendations

The NPS incorporates what it feels are best practices in teaching wargame design, data collection, and analysis.

The Naval Undersea Warfare Center
Overview

NUWC is one of two divisions whose mission is to serve as the Navy's full-spectrum research, development, test and evaluation, engineering, and fleet support center for submarine warfare systems and many other systems associated with the undersea battlespace.[6] As such, NUWC provides the technical and subject matter expertise required for the Navy to maintain its undersea superiority. This mission involves conceptualization, research, development, fielding, modernization, and maintenance of various systems.[7]

Wargaming at the Naval Undersea Warfare Center

NUWC utilizes wargaming to support and provide guidance to its overarching technical and research mission. NUWC's gaming capability is relatively new, but already supports several different games a year with a small staff of analysts. The wargaming program at NUWC may be characterized as support to training and education, support to concept development, and senior leadership engagement and strategic discussion.

The wargaming program at NUWC supports a broader community that provides coordinated engagement with the commander of the Fleet Forces Command and the commander of the Pacific Fleet to execute the integrated force development cycle. NUWC considers this cycle to be a variation on Peter Perla's cycle of research.[8] There are four lines of effort:[9]

- workforce development to increase the pool of analysts, engineers, and scientists familiar with wargaming and able to support the other three lines of effort
- concept development through collaborative wargaming efforts that bring personnel together from the ten Naval Sea Systems Command warfare centers

---

[5] Jeff Appleget and Fred Cameron, "Analytic Wargaming on the Rise," *Phalanx*, Vol. 48, No. 1, March 2015, pp. 28–32.

[6] Naval Sea Systems Command, "Warfare Centers: NUWC Newport Division," webpage, undated.

[7] Interview with NUWC staff, Newport, R.I., August 4, 2016.

[8] Perla, 2012, pp. 251–253.

[9] Email correspondence with NUWC staff, July 18, 2017.

- tactics, techniques, procedures, and concept of employment support to fleet stakeholders and the Warfighting Development Commands
- subject matter support to wargaming events at other commands such as Navy Warfare Development Command, ONA and the NWC; and organizations such as the Center for Strategy and Budgetary Analysis, the CNA, and RAND.

Some specific examples of NUWC's recent wargaming activities include the following:[10]

- Leading in design and development on the Chief of Naval Operation's Rapid Innovation Cell's Fleet Battle School game project to prototype a distributed, tablet-friendly game environment leveraging commercial design conventions and techniques
- gaming and workshop support to several of the Office of Naval Research technology exploration efforts
- tactics, techniques, and procedures development wargames for Submarine Force, Pacific Fleet and the Undersea Warfighting Development Center, including initial design consultation on the Low Resolution Tactical Simulation game
- development of virtual world capabilities to provide visualization of game-developed schemes of maneuver for tactical problems.

The wargaming program at NUWC emphasizes research and development objectives in its wargames (design, execution, and analysis), while leveraging technology to capture data and insights. The types of wargames can vary considerably. They range from manual, Kriegsspiel-style board-based games with four to eight players to larger hybrid, workshop/narrative games intended to provide additional structure to traditional brainstorming and facilitated discussion techniques.

Technology utilization at NUWC varies from acetate overlays on nautical charts adjudicated using Microsoft Excel spreadsheet models to advanced visualization techniques such as the VR tools described above. Integration of simple adjudication capabilities (through such spreadsheets) is in the planning stages. NUWC also coordinates requests to provide Naval Sea Systems Command Warfare Center data on naval technology in wargames. Efforts to expand this to data maintained by other warfare centers are in the initial stages.

Tools and Approaches

NUWC primarily relies on Fleet Battle School and Virtual Worlds software, narrative gaming, and matrix gaming. Partly because of the need to support geographically remote players, NUWC has invested considerable effort in building tools to enable

---

[10] Interview with NUWC staff, Newport, R.I., August 4, 2016; email correspondence with NUWC staff, July 18, 2017.

Table A.2
Naval Undersea Warfare Center Wargaming Tools and Approaches

Tool or Approach	Description	Usage
Matrix games	Means of structuring gameplay between multiple teams that enables players to weigh in on the likely outcome of other teams' actions.	Used as a substitute for seminar-style games, particularly when gaming emerging topics with limited expertise to inform adjudication.
Narrative games	Method for running seminar-style games aimed at developing innovative concepts.	Used for wargaming design, adjudication, and data collection.
Stand-alone computer network	Stand-alone networks for testing software and running games that are separate from usual government network approval processes.	Provides a space for new software to be used in game design, execution and analysis without posing a risk to the main system.
Virtual worlds	GOTS VR tool based on the open-source software OpenSimulator.[a]	Used for analysis, data collection, and visualization in Navy training and wargaming support.

[a] See OpenSimulator, homepage, undated.

gameplay over distance. Some of these games are based on rigidly assessed manual games, but others focus on less structured narrative gameplay.[11] A summary of NUWC tools and approaches is presented in Table A.2.

NUWC gamers we interviewed also stressed the importance of gaming CoPs in providing wargaming education. This includes both the Navy virtual CoP that has recently been stood up and broader-based organizations such as the MORS Wargaming CoP.[12]

Facilities

NUWC has several conference rooms available for games, as well as an innovation space that can host sessions in configurable rooms with whiteboards and stand-alone networks. The innovation space provides a space for experimenting with both in-person and virtual tools, and the goal is to experiment with tools that can support the flow of games. Equipment includes Apple, Linux, and Windows computers, large touch screen displays, Android and iOS tablets, Oculus Rift VR headsets, and three-dimensional printers to create game pieces or other objects.[13]

---

[11] Interview with NUWC staff, Newport, R.I., August 4, 2016.

[12] Currently, the Navy virtual CoP is accessible only with a Navy-issued Common Access Card, but efforts are underway to make it more accessible to the broader DoD gaming community.

[13] Interview with NUWC staff, Newport, R.I., August 4, 2016; email correspondence with NUWC staff, July 18, 2017.

Process and Skill Sets

As a relatively new gaming center, NUWC is currently building up processes and staff able to support diverse games. Process varies significantly with the demands of the specific game. Digital games require support from the Virtual Worlds team or programs like Fleet Battle School. This includes the skills needed to design the game itself, as well as those needed to use the digital tools effectively so that they augment, rather than derail, discussion.

Other games focus more heavily on manual game design and facilitation skills. NUWC is able to draw on a robust base of undersea warfare SMEs both from the broader staff and other Centers of Excellence. This is particularly critical for technical games, where these individuals can provide data, expertise, and analysis.

Key Best Practices and Recommendations

NUWC had a number of recommendations for the Marine Corps as it looks toward expanding the capabilities of its wargaming center. NUWC stressed that when building a new capability, it is important to start small and with an experimental attitude. Investing in flexibility and multipurpose tools allows for resilience when an idea does not work out and for experimentation to find the next helpful tool or process.[14] Facility and equipment recommendations included a space with multiple rooms; a stand-alone network; and computers with a variety of operating systems, as well as tablet and interface options.

NUWC staff also highlighted the importance of building up junior staff with an interest in wargaming who can then support efforts in the long term. In some cases, such staff may be formally assigned, but in others it may be helpful to cultivate skills outside the formal bounds of the organization to create a knowledge base about gaming in other areas of the organization.[15]

Naval War College Wargaming Department
Overview

The NWC has several missions: educating and develop future leaders; helping to define the future Navy and its roles and missions; supporting combat readiness; strengthening global maritime partnerships; promoting ethics and leadership through the force; contributing knowledge to shape effective decisions through the expertise of the John B. Hattendorf Center for Maritime Historical Research; and providing expertise and advice to the international legal community through the Stockton Center for the Study of International Law.[16]

---

[14] Interview with NUWC staff, Newport, R.I., August 4, 2016.

[15] Interview with NUWC staff, Newport, R.I., August 4, 2016.

[16] NWC, "Our Mission," webpage, undated b.

In keeping with these missions, it has six academic colleges and a number of PME programs. The NWC sees itself as "a place of original research on all questions relating to war and to statesmanship connected with war, or the prevention of war."[17] It also has an extensive history of wargaming.

Wargaming at the Naval War College

Located on the NWC's Newport, Rhode Island, campus, the Wargaming Department conducts research, analysis, gaming, and education. As part of the Center for Naval Warfare Studies, the Wargaming Department strives to prepare naval officers for future challenges and assist in shaping the future of the Navy.[18] Wargaming at the NWC is used to foster education in decisionmaking environments, building understanding for military and civilian leaders, generating and examining strategic and operational concepts, and providing operational insights.[19] NWC wargaming spans all categories of wargaming that we have defined for the purposes of this report. A varied repertoire of games is consistent with the long history of NWC wargaming and its continued preeminence among national security wargaming centers.

The NWC runs dozens of wargames a year. The games examine a wide number of topics, including space, cybersecurity, command and control, emerging technology, sea control, international maritime cooperation, and interagency coordination. Recently completed wargames at the NWC include the Global Title 10 Wargame Series for the Chief of Naval Operation, a command and control standardized task force construct game for the Pacific Fleet, and a deterrence and escalation game and review for the commander of U.S. Strategic Command.[20]

In addition to its operational and strategic wargames, the NWC conducts two Joint Military Operations capstone exercises annually, one for the intermediate-level students and one for senior-level students. Each exercise is an intensive event, involving roughly 300 students. Furthermore, to support the educational mission of the NWC, the Wargaming Department provides gaming support to various courses, departments, and organizations within the NWC. The Wargaming Department faculty teach courses on wargaming for both NWC students as an elective, and for external audiences as a weeklong wargaming professionals' course.

To address this kind of diversity in sponsors and objectives, the NWC employs a variety of approaches and tools within its wargaming program, which are further explained below.

---

17 NWC, "About U.S. Naval War College," webpage, undated a.

18 NWC, *War Gaming: United States Naval War College*, brochure, undated c.

19 NWC, undated c.

20 NWC, untitled c.

Tools and Approaches

Due to the vast array of topics for its wargames, the Wargaming Department utilizes several techniques, approaches, and tools. These include COTS, government PORs, and NWC-developed tools. Examples include technological enablers like the Joint Semi-Automated Forces (JSAF) and methodological processes like analysis of competing hypotheses (ACH). Nevertheless, the NWC emphasizes rigorous methods and analysis with all of its wargaming activities.[21] A summary of the NWC's primary tools and approaches for wargaming is presented in Table A.3. Other tools include i2 Analyst's Notebook and the SPSS. The Wargaming Department has also replaced SharePoint with Google Drive and Google Sites for knowledge management during wargames.

Facilities

Based in McCarty Little Hall, a 110,000-square-ft purpose-built wargaming facility, the NWC Wargaming Department has a 180-seat auditorium, a television studio, conference facilities, office and classroom space, and up to 22 reconfigurable gaming cells. In terms of capabilities, McCarty Little Hall maintains a state-of-the-art IT suite, which can support a full range of models and simulations, video teleconferencing (VTC), group collaboration systems, and distributed wide-area gaming over both unclassified and classified networks—all of which are intrinsic to the institution.[22]

The auditorium is equipped with projectors, classified and unclassified VTC capability, and the space to support a rehearsal of concept drill featuring a 25-ft-by-40-ft floor map. Meanwhile, the gaming cells can support both classified and unclassified events. The Joint Command Center is the largest of the game cells, featuring approximately 100 computer stations.[23] The Wargaming Department maintains seven separate networks at various classification levels, runs an extensive number of servers, and supports more than 400 thin client terminals.[24]

Process and Skill Sets

The Wargaming Department has published the *War Gamers' Handbook*, which documents many departmental processes and procedures for game design and management as they are run at NWC. The handbook reviews both the different roles necessary for game design, and discusses the department's preferred division of labor. Figure A.2 illustrates the department's project management process for wargames. Integral to this process is the initial sponsor tasking, as the design flows from the purpose. From there, development and playtesting are also critical. The handbook discusses a range

---

[21] Interview with NWC Wargaming Department staff, Newport, R.I., August 3, 2016.

[22] NWC, untitled c.

[23] NWC, untitled c.

[24] NWC, untitled c; interview with NWC Wargaming Department staff, Newport, R.I., August 3, 2016.

Table A.3
Naval War College Wargaming Tools and Approaches

Tool or Approach	Description	Usage
ACH	Analytic process to systematically enumerate and evaluate a complete set of hypotheses based on all available evidence.	Used for naval wargaming postgame analysis in deductive, hypothesis-driven gaming.
ATLAS.ti	Toolset to manage, extract, compare, explore, and reassemble meaningful pieces from large amounts of data in flexible yet systematic ways.	A primary tool for grounded theory and other qualitative analysis at the NWC.
Configurable gaming space	Large, configurable space for gaming.	Used as a venue for diverse sets of wargames.
FacilitatePro	Software that provides brainstorming, prioritizing, evaluating, surveying, and action planning tools to aid creativity and solve complex problems.	Used for in vivo coding in qualitative analysis.
Google Earth	A virtual globe and mapping tool that allows viewing of satellite imagery, maps, and terrain.	Used to increase immersion and situational awareness for players.
Google Drive and Google Sites	Allows for the creation of websites that act as a secure place to store, organize, share, and access information.	Used for information sharing and knowledge management during wargames.
JSAF	Computer-generated system that provides entity-level simulation of ground, air, and naval forces.	Used for wargame adjudication, visualization, and planning.
Large-format printers	Printers that allow a maximum width of 18–60 inches.	Used for high-quality prints of maps and game pieces, which are useful for increasing game immersion.
NWC web applications	Applications used to custom build graphical user interfaces (GUIs) and modules to support game execution and data collection tasks.	Used for Navy wargaming support.
Spreadsheet tools	Excel spreadsheets and macros.	Used for wargaming data analysis.
Stand-alone computer network	Networks for testing software and running games that are separate from usual government network approval processes.	Provides a space for new software to be used in game design, execution and analysis without posing a risk to the main system.
Videos/video studio	Ranging from a studio for live video production to camcorders and computers with editing software; several centers stress the utility of being able to produce "new" video to deliver game scenario briefings and injects for heightened impact.	Used for introduction briefings of the game scenario and scenario updates.

Figure A.2
The U.S. Naval War College Wargaming Project Management Process

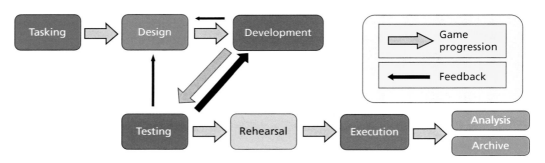

SOURCE: Burns, 2013, Figure 1.
NOTE: Adapted from DellaVolpe (2012), Logel (2012) depictions.

of design factors for consideration, such as the level of war, number of sides, players, adjudication, data collection, analysis, and practical aspects of event management.[25] The wargame process generally lasts nine months, from understanding the sponsor's objectives to execution. In analytic gaming, the postgame process may be as long as two months, depending on requirements and complexity.[26]

We were able to have a more detailed discussion about the software tool development process that the Wargaming Department uses to support its games. The Wargaming Department technology services prefer to have nine months for tool development: three months for initial development, three months for testing (including separate alpha and beta tests), and three months for other software preparations. A typical process may take five different people to develop and prepare the tool, and 30 people to do "destructive" testing of a system. Given the large number of simultaneous users anticipated for some of the department's wargaming tools, the destructive testing is meant to alert the technology services team to potential problems from a large number of users. Although each game can be different, "average" game support has several requirements. One is knowledge management of emails, requests for information, and other information and communication. A game may require dozens of visualization tools to handle order of battle, overlays, maps, time, and other visual aspects the game. Another dozen or so tools are needed to support the white team for tasks including adjudication, note taking, and ethnography.[27]

The Wargaming Department harnesses a diverse range of skill sets, including operational experience, design acumen, scenario development, quantitative and quali-

---

[25] Shawn Burns, ed., *War Gamers' Handbook: A Guide for Professional War Gamers*, Newport, R.I.: Naval War College, Wargaming Department, 2013.

[26] Email correspondence with NWC staff, August 2, 2017.

[27] Interview with NWC Wargaming Department staff, Newport, R.I., August 3, 2016.

tative analysis, and formal training in facilitating discussion. Wargaming Department staffing as of August 2017 consisted of 20 civilians, 15 officers, 40 enlisted members, and 30 contractors. This includes the technical staff discussed below.[28] A significant percentage of the department's faculty possess advanced degrees—including earned doctorates rather than natural degrees. The Wargaming Department finds that including these social science methods are critical to behavioral decisionmaking in gaming, since the preponderance of problems and data encountered are qualitative rather than quantitative.

The department also has technical staff to manage the configuration, maintenance, and development of physical and digital gaming spaces. At the time of our visit in August 2016, the department employed 35 technical staff and estimated that it spent $4 million–$4.5 million annually on operations and maintenance funding for these personnel. The department also noted that this did not cover research, development, testing, and evaluation funds for new tools, which it had to bring in through other means.[29]

Key Best Practices and Recommendations

The NWC Wargaming Department offered a number of recommendations for Marine Corps wargaming. Faculty and staff emphasized the importance of investing in flexible capabilities that enable a wide range of games. This goes both for personnel and facilities. In the case of personnel, the center recommended investing in a staff with a diverse range of skills that would enable a center to run a spate of different types of games, as well as experimenting with new tools and techniques based on sponsor demand.[30]

Physical and digital gaming spaces should also value flexibility by focusing on multipurpose reconfigurable space. Departmental personnel recommended against custom-built spaces that could not be reconfigured or repurposed for other uses. They also recommended stand-alone networks at both classified and unclassified levels to enable greater freedom to experiment with new tools. Given the increase in demand over time for events at higher levels of classification, personnel suggested that the Marine Corps consider including any wargaming facility with SAP/SCI requirements in its design concept.[31]

In terms of recommendations on tools, NWC wargaming faculty and staff noted that it was worth the Marine Corps having a discussion between a setup that depended on COTS with commercial support ("buy") versus one that depended on custom-developed tools ("make"). Departmental staff felt that certain commercial vendors

---

[28] Email correspondence with NWC staff, August 2, 2017.

[29] Interview with NWC Wargaming Department staff, Newport, R.I., August 3, 2016.

[30] Interview with NWC Wargaming Department staff, Newport, R.I., August 3, 2016.

[31] Interview with NWC Wargaming Department staff, Newport, R.I., August 3, 2016.

such as Adobe offered excellent support. If, on the other hand, the Marine Corps wanted to go with custom-developed tools, the NWC offered to share it hundreds of NWC-developed software applications for wargaming.[32]

Another issue that the NWC raised was personnel. The Wargaming Department consists mostly of government civilians and military staff, and proportionally few contractors, and the department suggested that the Marine Corps contract staff to provide labor, rather than a certain number of games, in order to give the Marine Corps maximum flexibility in how to assign personnel.

One final recommendation from the NWC was that the Marine Corps should organize and manage its wargaming to ensure intellectual autonomy. As a military university covered by the American Association of University Professors standards on academic freedom (which are codified in the NWC *Faculty Handbook*), allowing for objective independent conclusions free from sponsor bias is essential to examine issues with honesty and integrity.

The Office of Naval Intelligence
Overview
As a core element of the Navy's information warfare community, ONI is tasked with gaining and holding a decisive information advantage over America's potential adversaries in the maritime domain.[33] Thus, ONI must provide decisionmakers with relevant maritime intelligence, which includes adversary maritime warfare operations, scientific and technical capabilities, weapons proliferation, transnational threats, counternarcotics efforts, and the global maritime environment. The staff at ONI consists of a mixture of active duty and reserve naval personnel and civilians.[34]

Wargaming at the Office of Naval Intelligence
ONI wargames give junior analysts experiential understanding of military roles and missions and how strategic objectives, national resources, strategy, and force structure must all align. ONI uses wargaming mostly for training purposes, and thus falls under our category of wargames for training and education.[35]

Tools and Approaches
ONI's primary wargaming tool is the Simulation-Based Analyst Training (SimBAT) program. SimBAT is a tabletop game based on commercial board games, and its goal is to both expose junior analysts to the dynamism and complexity of warfare and intelligence analysis and provide ONI analysts and managers a training experience closely aligned to ONI analytic standards.

---

[32] Interview with NWC Wargaming Department staff, Newport, R.I., August 3, 2016.

[33] ONI, "Our Mission," webpage, undated.

[34] Email from ONI staff, September 19, 2017.

[35] Telephone interview with ONI staff, October 14, 2016.

SimBAT courses introduce students to the basics of the Joint Planning Process through a series of structured analytic exercises and commercial board games adapted to meet professional training requirements. Scenarios typically involve World War II so as to teach some history while demanding that students plan and implement their strategy for global joint warfare. The facilitator helps students experience the what, how, and why of the strategic military decision process. Students are able to manage air and sea assets and to think through trade-offs in procurement.[36]

From 2008 to 2011, SimBAT courses ran three to five days in length. Since sequestration, the program has run as a two-day module in ONI's introductory training course. ONI's SimBAT program employs modified commercial board games as reviewed in Table A.4. ONI identifies cost-effectiveness as one of the advantages of board games.[37]

Facilities

ONI does not have dedicated facilities for wargaming but is able to access several rooms for events. ONI uses rooms for team collaboration during the planning process, and another plenary room for both teams to participate across a common game board. Due to the minimal demands of board game–based wargames, ONI's wargaming program operates on minimal resources.[38]

Process and Skill Sets

SimBAT courses consist of experiential learning exercises followed by an in-depth after action review. The wargames are the "lab" part of larger training courses that combine

Table A.4
Office of Naval Intelligence Wargaming Tools and Approaches

Tool or Approach	Description	Usage
SimBAT	Program within ONI that serves both training and analysis objectives.	Used for junior analyst training by providing a multidimensional, "all-sensory" training model to junior analysts that familiarizes junior analysts with the staff planning process and strategic decisionmaking.
SimBAT materials: modified commercial board games	Use of commercially available games for educational or analytical use; modification is generally needed to make rule sets more accessible and create a playable game to address the educational requirement.	Used for education and experiential learning on topics including COA development.

---

36 Telephone interview with ONI staff, October 14, 2016.

37 Telephone interview with ONI staff, October 14, 2016.

38 Telephone interview with ONI staff, October 14, 2016.

briefings with pregame exercises designed to help students understand the military factors being modeled in the simulations. SimBAT courses feature a "red hat" phase of intelligence assessment, followed by strategic planning, the wargame, and the after action review of the COA selection. This process allows junior analysts to compare and contrast their decisions relative to the adversary within the wargame. To encourage dialogue and analysis, the wargame employs Socratic elicitation and structured analytic techniques (SATs) consistent with the Joint Planning Process. The intelligence training and decisionmaking focus on uncertainty over time, space, and enemy intentions and capabilities. Junior analysts must work through procurement trade-offs, as the ONI mission involves multiple and diverse air, sea, and land threats.

ONI wargames present a sense of immersion in order to promote learning and retention, and they aim for a moderate level of stress for students in the wargames. After action reviews help students compare their views of the adversary before and after the game, rebalance after playing competing sides, and step back to assess performance.

SimBAT wargame management follows the NWC model, whereby a white cell adjudicates actions by red and blue cells. Games always involve rigid Kriegsspiel-style gameplay, with strict rules governing reconnaissance, movement and combat.

As for personnel, SimBAT requires at least one multicompetent program manager, course designer, instructor, and facilitator familiar with various elicitation techniques to improve student engagement and learning. One individual can run the simplest games where the goal is to emphasize intelligence estimation strategy and force structure design. For more complex courses and simulations, up to five white-cell facilitators may be required as adjudicators and embeds in red and blue cells. White-cell members can typically be trained in their duties in a day or two, but experience and aptitude for coaching and mentoring are necessary.[39]

Key Best Practices and Recommendations

ONI believes that tabletop wargames that are based on board games are cost-effective and easily understood by students. ONI finds these games better designed for high levels of technical validity and better suited to the classroom than are more expensive computer wargames designed for entertainment purposes. The latest format of the SimBAT program only requires a single individual to run it.

In ONI's assessment, recent generations of junior analysts bring much less basic military background knowledge to their careers than was the case a few decades ago. They need more training in the elementary fundaments of strategic situation assessment, force structure, and the roles and missions of land, naval, and air force components. Basic commercial board wargames provide excellent raw material around which courses can be designed.

For training value, however, SimBAT course sequence is critical. The wargame lab builds on the team-based pregame analytic exercises in which the student teams

---

[39] Telephone interview with ONI staff, October 14, 2016.

(blue and red cells) come to a collective understanding of the military factors under consideration. Postgame after action reviews then assist students in converting impressions from the highly dynamic experience into learning takeaways that are coherent and memorable and that deepen their intuitive understanding of the dynamics of warfare and staff planning.[40]

U.S. Army Wargaming Centers

The U.S. Army also has several organizations that wargame, and the organizations listed here hardly encompass the complete set of Army wargaming activities. However, between the CAA, the CGSC, TRAC, and USAWC, these organizations represent a variety of Army wargaming activities with insights for the Marine Corps. The frequent collaboration between the Army and the Marine Corps also makes a discussion about Army wargaming a natural part of the conversation about Marine Corps wargaming. We also include UFMCS because of the existing connection between UFMCS and the Marine Corps, and the emerging ties between red teaming and wargaming.

Army Command and General Staff College

Overview

The CGSC at Fort Leavenworth, Kansas, identifies itself as an organization that "educates, trains and develops leaders for Unified Land Operations in a Joint, Interagency, Intergovernmental, and Multinational operational environment; and advances the art and science of the profession of arms in support of Army operational requirements." It has fourteen academic programs, including three that are able to grant graduate degrees. The CGSC is part of Army University and was conceived as the School of Application for Infantry and Cavalry in 1881.[41]

Wargaming at the Command and General Staff College

The Directorate of Simulation Education (DSE) is the simulation and gaming center supporting educational programs at the CGSC. The DSE also supports the School of Advanced Military Studies. DSE wargaming emphasizes support to training and education. Its main focus is tactical and operational gaming and simulations that support the college's educational mission. The division uses both computer and manual games to encourage classroom discussion, reinforce learning, and stress decisionmaking for its students. The DSE is able to support 20 student sections at a time. It is also involved in wargaming to support operational decisions and plans. Wargaming for Army sponsors such as the 3rd Army and the 35th Infantry Division helps to maintain the relevance in operational areas.[42]

---

[40] Telephone interview with ONI staff, October 14, 2016.

[41] U.S. Army Combined Arms Center, "Mission, Vision, Priorities, Principles, & College-Level Learning Outcomes: About the Command and General Staff College," webpage, undated a.

[42] Interview with CGSC staff, Fort Leavenworth, Kan., August 20, 2016.

Tools and Approaches

The DSE relies predominantly on commercial board games, modified commercial board games, and custom PC games. Digital games included a brigade-level game titled *Decisive Action: Future Force* used in the force management class; *Forward into Battle*, which helps students learn to move forces forward to assembly areas while protecting and fueling them; and *The Grand Offensive*, a game of decisionmaking in World War I.

The DSE is also notable for its use of board games. For example, it uses the original Kriegsspiel board game from 1824 Prussia to teach CGSC students mission command and how to write mission orders. Drive on Paris is a hex-and-counter game that the DSE uses in a CGSC logistics elective. The staff noted several advantages to commercial board games and modified commercial board games, including ease of use as computer networks were becoming harder to use due to security restrictions, the benefits of group dynamics, the ease of prototyping, and the relatively minimal cost. The organization also reported finding a reference library of commercial board games and a stand-alone computer network to be useful tools.[43] A summary of the CGSC's primary tools and approaches for wargaming is presented in Table A.5.

Facilities

The CGSC is located at Fort Leavenworth as part of the U.S. Army Combined Arms Center. The Simulations Division has one dedicated classroom for its activities.

Process and Skill Sets

The DSE had a small staff of about seven individuals—a mix of civilians, military personnel, and contractors—at the time of RAND's visit. The directorate's lead is a lieutenant colonel simulation operations officer (Functional Area FA57) with a background in M&S and knowledge management.[44] The DSE appears to be one of the few wargaming organizations we contacted that had a specified military occupational specialty requirement for its director. Staff skill sets at the time of RAND's visit included computer programming, scenario development, artwork, and game facilitation. The Simulations Division did not find a lack of prior military knowledge to be a disadvantage in its hires, and instead preferred to find individuals with hobby gaming experience—particularly those with a penchant for gaming activities focused on supporting the game and players, such as game mastering or scenario creation.[45]

The DSE works with CGSC educators to support classroom objectives and produces about one new game a year. Its process for approaching new games is to work with the educators to understand the educational purpose, what decisions the students

---

[43] Interview with CGSC staff, Fort Leavenworth, Kan., August 20, 2016.

[44] U.S. Army Modeling and Simulation Office, "Military Program—FA57," webpage, undated.

[45] Interview with CGSC staff, Fort Leavenworth, Kan., August 20, 2016.

Table A.5
Army Command and General Staff College Wargaming Tools and Approaches

Tool or Approach	Description	Usage
Custom PC games	Computer-based tools modeling tactical decisionmaking in such conflicts as World War I.	Used to illustrate a relatively specific series of military choices in a classroom context rather than providing detailed or historically accurate analysis.
Gaming reference library	Library of commercial board games and game design books for reference.	Used for reference and inspiration when creating additional games.
Modified commercial board games	Use of commercially available games for educational or analytical use. For all but the simplest games, modification is generally needed to make rule sets more accessible, and create a playable game to address the educational requirement.	Used for education and experiential learning on topics including COA development, principles of counterinsurgency (COIN), and ethics and decisionmaking.
Stand-alone network	Separate computer network to install games and tools.	Used to experiment with new games and tools.

would need to make in the game in order to meet that purpose, and then to create the sets of interactions that create opportunities for those decisions.[46]

Key Best Practices and Recommendations

The DSE views wargames as helpful learning tools, especially in conjunction with broader military education. Tools that it recommended included commercial board games, modified commercial board games, commercial PC games, a gaming reference library, and a stand-alone network.

The DSE feels that an educational partnership with service institutions of higher learning could prove fruitful for other wargaming centers. Its perspective is that the arrangement exposes military personnel to the benefits of wargames and poses the opportunity to build and sustain long-term interinstitutional relationships.[47]

The Center for Strategic Leadership at the U.S. Army War College

Overview

Based in Carlisle, Pennsylvania, USAWC serves as the senior educational institution of the U.S. Army. Its goal is to produce national security experts—both military and civilian—in strategic thinking and the role of ground forces in national security. USAWC includes a number of institutions. The School of Strategic Landpower provides resident and distant joint professional military education (JPME), with an

---

46 Interview with CGSC staff, Fort Leavenworth, Kan., August 20, 2016.

47 Interview with CGSC staff, Fort Leavenworth, Kan., August 20, 2016.

emphasis on the strategic level. CSL activities include strategic pol-mil simulation; support to Army and joint exercises, research, and analysis; and interagency training and education. The Strategic Studies Institute engages in research, conferences, publishing, and analysis for the Army and Joint Staff. Other elements of the USAWC, including the Army Heritage and Education Center, the Peacekeeping and Stability Operations Institute, and the School of Strategic Landpower, provide subject matter expertise for the wargaming department's games.[48]

Wargaming at the U.S. Army War College

USAWC's wargaming capacity resides predominantly within the CSL's Department of Strategic Wargaming (DSW). The DSW's mission is to use wargames, exercises, and simulations to assist in the development of strategic leaders and advisers, foster strategic innovation, improve strategic planning, and advance understanding of strategic issues for the Army, the Joint Force, and the nation.[49] The DSW is divided into the Wargame Operations Division, the Strategic Simulation Division, and the Strategic Assessments and Operations Research Division.[50] We spoke with members of the DSW about their wargaming tools, approaches, and practices.

USAWC wargaming supports training and education and wargaming to support operational decisions and plans. Educational wargaming is a notable part of its wargaming, particularly given the college's mission.[51] Many wargames focus on the strategic level and illustrate or aid classroom discussion.

USAWC does not have a standard format for its wargames but combines different techniques and approaches for any given customer's wargames.[52] Overall, USAWC is moving away from seminar-style games (due to their limitations) to games with red cells and more competition. The intent is to move from a discussion of what a side in a situation should do to taking action in a game. Game adjudication sometimes consists of two or three people who determine the result (expert or free adjudication), sometimes involves dice rolls (rigid adjudication), and sometimes involves a blend of free and rigid adjudication (semirigid adjudication).[53] Rigid adjudication games are often modified from commercial games, as the center has found that relatively little change is needed to support educational objective. Figure A.3 shows a matrix game, Kaliningrad, developed by the DSW for students in the European Regional Studies Program. Kaliningrad has subsequently been modified and reused several times to support team

---

[48] Email from USAWC staff, September 19, 2017.

[49] Email from USAWC staff, September 19, 2017.

[50] USAWC, "Department of Strategic Wargaming (DSW)," webpage, undated c.

[51] Telephone conversation with USAWC staff, November 18, 2016.

[52] Telephone conversation with USAWC staff, November 18, 2016.

[53] Telephone conversation with USAWC staff, November 18, 2016.

Figure A.3
The U.S. Army War College Kaliningrad Matrix Game

SOURCE: USAWC, "January–June, 2017: International Fellows (IFs) Matrix-Game," webpage, undated d.

building and experiential learning by international fellows at USAWC, as explained below.[54]

USAWC also wargames in support of operational decisions and plans as part of its broader research and analysis support to the Army and Joint Force. An example of this is wargaming that looks at the Army mobilization process. Because the intent of these wargames is to inform decisionmakers about a particular problem, the emphasis is not on participant learning. Although USAWC itself and the Army are two key customers for USAWC wargames, it also has a history of providing wargame support to joint task forces and CCMDs. USAWC also participates in joint wargaming efforts with the other service war colleges, such as the Joint Land, Aerospace and Sea Strategic Exercise (JLASS).[55]

---

[54] USAWC, "The Center for Strategic Leadership," webpage, undated b.

[55] Telephone conversation with USAWC staff, November 18, 2016; email from USAWC staff, September 19, 2017.

Tools and Approaches

Generally, USAWC conducts matrix- or seminar-style wargames. These types of wargames are best suited for the classroom setting, where discussion is more important than providing a specific concrete answer or solution to a problem. A summary of USAWC's major wargaming tools and approaches is given in Table A.6. Other tools include process maps (similar to concept maps), digital and physical maps, sticky notes, smart boards, and even Lego pieces.[56] One recent addition to USAWC's toolkit is red teaming, which USAWC staff will acquire through the UFMCS six-week red teaming course.[57]

Facilities

Collins Hall, the home of the CSL, is a purpose-built facility for wargaming. The second floor of Collins Hall consists of two distinct wargaming centers with up to 20 distinct spaces that can be reconfigured and combined to form larger gaming rooms as required. Each room is equipped with AV equipment, as well as computers and associated peripheral devices. Collins Hall can host up to five simultaneous networks,

Table A.6
Army War College Wargaming Tools and Approaches

Tool or Approach	Description	Usage
The Applied Critical Thinking Handbook (formerly the Red Teaming Handbook)	Handbook of structured group methods aimed at reducing bias, mitigating groupthink, and encouraging critical thinking.	Used for framing questions, generating approaches, and evaluating potential approaches.
Large-format printer	Large or wide format printers are generally accepted as any printers that allow a maximum width of 18–60 inches.	Used for printing maps, posters, and various visualization aids for use during games.
Matrix games	Means of structuring gameplay between multiple teams that enables players to weigh in on the likely outcome of other teams' actions.	Used as a substitute for seminar-style games, particularly when gaming emerging topics with limited expertise to inform adjudication.
Modified commercial board games	Use of commercially available games for educational or analytical use.	Used for education and experiential learning on topics.
Seminar	Structured discussions on specific scenarios or problem sets, leveraging subject matter expertise.	Used for exploring strategic and operational impacts of a specific vignette and for exploring new ideas, concepts, and/or intangibles at the operational/strategic level.
Spreadsheet tools	Excel spreadsheets and macros.	Data capture and analysis.

---

[56] Telephone conversation with USAWC staff, November 18, 2016.

[57] Telephone conversation with USAWC staff, November 18, 2016.

including commercial unclassified internet, NIPRNet, SIPRNet, and stand-alone classified and unclassified exercise networks. Collins Hall has an 80-person executive conference room used for plenary sessions during wargames, as well as numerous smaller conference rooms, VTC suites, and multipurpose rooms. In addition, USAWC also uses classrooms and other multipurpose spaces outside Collins Hall.[58]

Process and Skill Sets

USAWC personnel involved in wargames include strategists, intelligence analysts, OR analysts, and professors.[59] USAWC finds OR analysts to be good at documenting the process of a wargame, understanding how parts interact, and translating processes into a wargame. Strategic and intelligence analysts are helpful in understanding the significance of decisions or events in the game and when something may be important to record.[60]

Data collection should be an analytical plan based on the game purpose and objectives. Whether data is qualitative or quantitative depends on the nature of the game. However, it tends to be fairly straightforward. Data collection is often accomplished by writing down information on a board or piece of paper and later entering the data into a spreadsheet. USAWC often records its sessions and listens for how people made decisions. However, USAWC destroys these recordings afterward because the games are not for attribution. Similarly, any transcripts from these recordings are also destroyed after use. Analysis of the data can involve simple statistics, pattern matching, and description of the raw game results.[61]

Key Best Practices and Recommendations

USAWC feels that educational wargames can provide an invaluable tool for classrooms and their respective students. Wargames allow students to formulate and experiment with different and creative decisionmaking. USAWC has tried to shift wargaming away from seminar-style discussions to approaches involving more structure and formal adjudication.[62]

Center for Army Analysis

Overview

The CAA conducts analyses across the spectrum of conflict in a joint, interagency, intergovernmental, and multinational context to inform critical senior-level decisions for current and future national security issues.[63] The CAA's core competencies include

---

[58] Email from USAWC staff, September 19, 2017.

[59] USAWC, undated c.

[60] Telephone conversation with USAWC staff, November 18, 2016.

[61] Telephone conversation with USAWC staff, November 18, 2016.

[62] Telephone conversation with USAWC staff, November 18, 2016.

[63] CAA, homepage, undated.

campaign analysis, strategic and operational assessments, deployed analytic support with reach-back, operational and institutional capability analyses, force structure analyses, M&S policy and strategy, and M&S workforce development. The CAA is located at Fort Belvoir, Virginia.

Wargaming at the Center for Army Analysis

The CAA conducts several types of wargames of varying sizes for different sponsors across the Army and Joint Force, and ranges from examining a set of COAs to exploring future warfighting capabilities. CAA gaming includes wargaming to support concept development, wargaming to support capability development and analysis, and wargaming to support operational decisions and plans. It conducts wargames at the service, CCMD, and joint levels. Overall, the CAA conducts between six and nine games a year. As a predominantly analytic organization, the CAA combines wargaming and analysis in several areas.

CAA wargaming around the joint technologies in the Long Range Research and Development Plan appears to involve both wargaming to support concept development and wargaming to support capability development and analysis. The Long Range Research and Development Plan is an effort at the OSD for Acquisition, Technology and Logistics to identify technology investments with the potential to affect future U.S. technological superiority.[64] The Army Capabilities Integration Center (ARCIC), the CAA, and TRAC have been involved in down-selecting and wargaming potential future technologies for ground combat, using the "best guess" of the service war colleges and the research and development community on what those technologies might look like. ARCIC, the CAA, and TRAC are then able to take wargaming results and conduct additional, more detailed modeling.[65]

Other support capability development and analysis encompasses CAA wargames around SSA products; these include scenarios, CONOPS, forces, and baselines, and they support department-level deliberations on PPBES matters—including force sizing, shaping, and capability development.[66] The CAA has wargamed with SSA products to test whether the CONOPS and forces associated with a scenario are sufficient. It has also wargamed SSA products with the joint community to create a consensus on how events would play out in order to enable additional modeling and analysis to support the PPBES, and it has wargamed to produce COAs and alternative COAs within the framework of SSA products.[67]

---

[64] Frank Kendall, Under Secretary of Defense for Acquisition, Technology and Logistics, "Long Range Research and Development Plan (LRRDP) Direction and Tasking," memorandum, Washington, D.C., October 29, 2014.

[65] Interview with CAA wargamers, Fort Belvoir, Va., December 12, 2016. The Long Range Research and Development Plan ground combat wargaming also included Marines.

[66] DoD, *Support for Strategic Analysis (SSA)*, Washington, D.C.: Department of Defense Directive DoDD 8260.05, July 7, 2011.

[67] Interview with CAA wargamers, Fort Belvoir, Va., December 12, 2016.

The CAA also assists CCMDs through wargames to support operational decisions and plans. It conducts wargames in theater to assist European Command and Pacific Command in testing their OPLANs. The CAA uses the CCMDs' existing wargame framework and also tests a different COA as a teaching tool.[68]

Tools and Approaches

The principal tool utilized by the CAA is JWAM, a joint, operational-level wargaming system that is manual and computer-assisted. JWAM involves two or more opposing teams engaged in free play where aggregated tactical outcomes form the wargames' operational-level insights. The model is divided into 13 distinct steps. It stresses Phase II and Phase III of operational campaigns, with 24- or 72-hour time steps.[69]

The 13 JWAM steps are as follows:

1. Determine Weather
2. Cyber/Space/EW [Electronic Warfare] Operations
3. ISR [Intelligence, Surveillance, and Reconnaissance] Operations
4. Integrated Air Defense System Allocation
5. Strategic Strike Missions
6. Determine Air Superiority
7. Strategic Deployment
8. Logistical Sufficiency Check
9. Naval Combat
10. Tactical Deep Strike Missions
11. Ground Combat
12. Post Combat
13. Post-turn Hot Wash[70]

JWAM represents a decade's worth of effort and adaption. Originally the model was heavily ground-centric and based on the wargame process used by the 1st Cavalry Division Plans Shop. Since then JWAM has been modified to better reflect the Joint Force and the capabilities of other services. The CAA uses Microsoft Access for the JWAM "battle tracker" and Excel for game analysis. OSD Acquisition, Technology and Logistics; OSD CAPE; and other organizations are currently involved in efforts to further digitize JWAM.

---

[68] Interview with CAA wargamers, Fort Belvoir, Va., December 12, 2016.

[69] Daniel P. Mahoney, *The Joint Wargaming Analysis Model*, version 8.1, Fort Belvoir, Va.: Center for Army Analysis, November 2016b, pp. 1–2.

[70] Daniel P. Mahoney, "Center for Army Analysis Wargame Analysis Model (C-WAM)," PowerPoint presentation for the Military Operations Research Society Wargaming Community of Practice, Arlington, Va., April 20, 2016a, p. 7.

Table A.7
Center for Army Analysis Wargaming Tools and Approaches

Tool or Approach	Description	Usage
Google Earth	A virtual globe and mapping tool that allows viewing of satellite imagery, maps, and terrain.	Can be integrated into a wargaming center's common operational picture (COP) for use by players, used during presentations as a visual aid, or as a source of printed high-resolution satellite imagery for game boards or other visual aids.
Large-format printers	Large or wide format printers are generally accepted as any printers that allow a maximum width of 18–60 inches.	Used for high-quality prints of maps and game pieces, which are useful for increasing game immersion.
JWAM	A manual, computer-aided, time-step, human-in-the-loop, force-on-force simulation methodology developed over ten years at the CAA.	Used to evaluate and compare COAs to support CCMD operational plans and for defense planning scenario development.
Spreadsheet tools	Excel spreadsheets and macros.	Data capture and analysis.

The CAA's main wargaming tools are listed in Table A.7. The CAA uses large, printed images from Google Earth and spreadsheet tools to support JWAM wargames.

Facilities

The CAA has several dedicated wargaming map rooms, and also uses common conference space certified as a temporary Sensitive Compartmented Information Facility for wargames. CAA also conducts a number of wargames at CCMD and other locations.

Process and Skill Sets

The three major steps to conducting CAA wargames are plan, execute, and postprocess. The entire process takes about three months; this includes two weeks to identify the red and blue teams, run the planning process, present briefings, conduct the mission analysis, and develop the most likely and most dangerous COAs. A typical JWAM game takes two weeks to run. The postprocess phase includes processing of notes and insights. Additionally, wargames that are meant to feed into analysis using M&S tools require that output be captured primarily as quantitative data.

The CAA's wargaming component at the time of our interviews in 2016 included 26 people, of whom eight were campaign analysts. Due to an ongoing hiring freeze, the CAA was at the time unable to expand its wargaming capacity. One skilled position that the CAA identified as particularly hard to fill was that of coders who understood both wargaming and analysis.

In terms of training or skills needed to run JWAM, the CAA noted that there was no formal training available. Instead, the CAA estimated that six months of practice with JWAM would make someone adept and felt that campaign analysts were ideal as JWAM gamers.

Key Best Practices and Recommendations

One key takeaway from the CAA was the advantage of keeping JWAM a largely manual, rather than computerized, process. This means that the equipment used to run JWAM consists of paper, pasteboard, dice, whiteboards, cork boards, and Microsoft Office tools such as Access and Excel.

Several times during its adaptation, the CAA considered modernizing JWAM through incorporating lightboards or DoD-developed computer wargaming tools such as SWIFT or VAST. However, the CAA purposefully decided to leave JWAM as a manual and paper-based simulation because the wargamers felt that the low-tech aspect of JWAM enhanced reliability when they needed to travel and also allowed for greater flexibility in future adaptations. Not having to secure access points for tools such as SWIFT, particularly at the Top Secret level and while traveling, was another perceived benefit of a manual system.

Another reason for staying manual was the concern that automating the process (from three-day turns) would mean moving participants through too quickly. CAA wargamers expressed the view that automated turns would lead analysts and participants to brush aside alternative COAs, branches, and sequels and skip over many conscious decisions that would otherwise surface during the game. The CAA continues to look for opportunities for wargaming automation that will enhance efficiency while maintaining the benefits of wargaming.

CAA wargamers also discussed what they felt were the advantages of a wargaming process such as JWAM over M&S-centered analysis. One thought was that JWAM results benefited from the direct input of service operators participating in wargaming, as compared with analysts modeling results from computer models such as the Joint Integrated Contingency Model or the Synthetic Theater Operations Research Model. While ideally the CAA or any analytic organization would triangulate results from the Joint Integrated Contingency Model, JWAM, and the Synthetic Theater Operations Research Model, JWAM results were considered especially valuable because of the expertise of wargame participants from outside the CAA.

The Research and Analysis Center

Overview

TRAC is part of Army Future Command. At the time of our interviews, it provided support to TRADOC and had a specific mission to produce relevant and credible operations analysis to inform Army decisions. The write-up herein reflects its mission at that time.

TRAC serves as the principal analytical organization of TRADOC, while unaligned with TRADOC proponents. TRAC provides centralized leadership and management of analysis for combat, training, and doctrinal developments. It conducts studies and analyses for TRADOC and Headquarters, Department of the Army; conducts studies of the integrated battlefield related to doctrine, organization, training, materiel, personnel, and leadership; designs and develops M&S for capabilities development; participates in technical exchange programs at the national and international levels; provides analytical support to ARCIC, the Centers of Excellence, and schools; directs research related to methods, models, and analysis; establishes, maintains, and manages the databases, scenarios, models, and wargaming tools required to support analyses and studies; and reviews and ensures, as directed, the quality of TRADOC studies before their approval.

The TRAC Mission Essential Tasks are as follows:

- Conducting the studies that inform key decisions made by Army, Joint Force, and TRADOC leaders.
- Leading analysis of current operations.
- Developing and maintaining the scenarios to underpin Army concepts and requirements.
- Developing, configuration managing, and applying verified and validated models and simulations.
- Researching, developing, and sharing new analytic methods and modeling.

TRAC has four locations: TRAC Fort Lee, New Jersey; TRAC Fort Leavenworth, Kansas (TRAC-FLVN); TRAC Monterey, California; and TRAC White Sands Missile Range, New Mexico (TRAC-WSMR). TRAC Headquarters is colocated at Fort Leavenworth with TRAC-FLVN. TRAC-FLVN and TRAC-WSMR constitute the largest parts of the center. Discussions about TRAC's wargaming focused primarily on TRAC-FLVN activities.

Wargaming at the Research and Analysis Center

Gaming at TRAC involves wargaming to support combat development and analysis, wargaming to support training and education, and wargaming to support concept development. TRAC-FLVN and TRAC-WSMR together run six or fewer wargames a year, most of which are map exercises (MAPEXes). Most of the wargaming within TRAC-FLVN is conducted by the Scenarios Group, which builds operational scenarios for Army analyses of alternatives. TRAC internally classifies its wargames/activities into five categories: workshop/SME roundtable, MAPEX, seminar, board game or manual game, and simulation- or tool-supported game. One of TRAC-FLVN's current major wargaming activities is writing a code of best practices for wargaming at TRAC.

TRAC utilizes wargames to support its Mission Essential Tasks; acquisition and programmatic analyses such as the analyses of alternatives, concepts and requirements development; and TRADOC standard scenario development. Less frequently, TRAC uses wargames to support current operations, experimentation, training, and education.

TRAC-FLVN and TRAC-WSMR together run or support approximately 12–18 wargames a year. Many of these wargames are MAPEXes supporting scenario development. The Scenarios and Data Directorate at TRAC-FLVN conducts most scenario development MAPEXes, which produce TRADOC standard scenarios required to support Army and joint Analysis of Alternatives, other acquisition-related efforts, and operational testing. Most other TRAC-led and TRAC-supported wargames are SME workshops, roundtables, or MAPEX variants that support studies through requirements development, CONOPS analyses, and the like. TRAC is currently developing a TRAC Wargaming Code of Best Practices.

Tools and Approaches

TRAC applies a range of OR methods, models, and tools, and it classifies "wargaming" as a method, model, and tool to support decision analyses. TRAC also utilizes large combat simulations to support analyses. Combat simulations that do not use humans in the loop during simulation execution are not considered wargames; such combat simulations include Advanced Warfighting Simulation (AWARS) and the Combined Arms Analysis Tool.

TRAC classifies its wargames into five categories or types: workshop or SME roundtable, MAPEX, seminar, board game or manual game, and simulation- or tool-supported game. Table A.8 summarizes TRAC wargaming tools and approaches.

Process and Skill Sets

Skilled personnel that TRAC employs in its wargames include OR and operations systems analysts, uniformed officers experienced in Army operations, and additional topical SMEs as required. TRAC teaches wargame design, execution, and analysis internally and draws from Army personnel to assist in running wargames.

TRAC wargames typically require smaller resource footprints (time and personnel) than the large combat simulations. Wargaming at TRAC has substantially increased in demand and use over the past several years, and TRAC expects this increased demand to continue.

TRAC's standard wargaming process involves four steps: planning, preparation, execution, and postgame. Regardless of the tools or approaches used, TRAC strives to emphasize uncertainty and human decisionmaking in its wargames. Typically, the results of TRAC wargames are used as inputs for other analysis or concept development within the Army. Figure A.4 illustrates the wargame design and development process at TRAC. It serves as the standard process and framework for TRAC wargames.

Facilities

TRAC does not have facilities specifically for wargames. It uses conference room spaces, but does have some spaces that can be configured to rapidly support wargame applications. TRAC does have workspace facilities and laboratories dedicated to its major combat simulations (and AWARS and the Combined Arms Analysis Tool).

Table A.8
The Research and Analysis Center Wargaming Tools and Approaches

Tool or Approach	Description	Usage
Workshop/SME roundtable	A structured discussion regarding system applications, CONOPS, etc., with technical or adversary SMEs providing expertise to inform and challenge applications and CONOPS.	Often used as a preliminary to a more formal MAPEX. Provides forum to better characterize the problems the study is addressing, the conditions under which those problems exist, and the attributes of potential solutions.
MAPEX	The planning of forces or a scheme of maneuver, development of a synchronization matrix, and/or COA development as part of the military decisionmaking process.	Used for exploring the tactical impacts of a specific vignette. TRAC uses MAPEXes extensively during scenario development and studies.
Seminar	A structured discussion on a specific scenario or problem set, leveraging subject matter expertise.	Used for exploring strategic and operational impacts of a specific vignette and for exploring new ideas, concepts, and/or intangibles at the operational/strategic level (Not used very often at TRAC.)
Boardgame—SSR: Mindanao	A cooperative, multisided leadership board game that covers shape and deterrence operations in the present day Philippines.	Used to inform students and staffs on the complexities associated with Phase 0 and Phase 1 operations and for practicing leadership attributes and competencies as outlined in Army doctrine.
Board game— commercial	Similar to MAPEX, but using a different medium.	Used for education and professional development by examining a historical battle or the practical application of taught material or doctrine.
Simulation-supported VAST	PC-based software tool using smart-board technology. Turn-based model.	Used as a visualization and adjudication support tool for MAPEXes. Digitally records unit movement and unit states.
One Semi-Automated Forces (OneSAF)[a]	Interactive, turn-based, human-in-the-loop constructive, stochastic, simulation.	Mainly used for adjudication, visualization, and data collection. (Currently used less frequently than other approaches at TRAC.)

[a] Because OneSAF is a less frequently used approach, it is not included in Appendix B.

Key Best Practices and Recommendations

TRAC concentrates on providing valuable insights through its wargames with high fidelity in results for further analysis. Overall, TRAC advocates that all of its wargames must

- meet analysis output requirements as defined with the study's data collection management plan

Figure A.4
The Wargame Design and Development Process

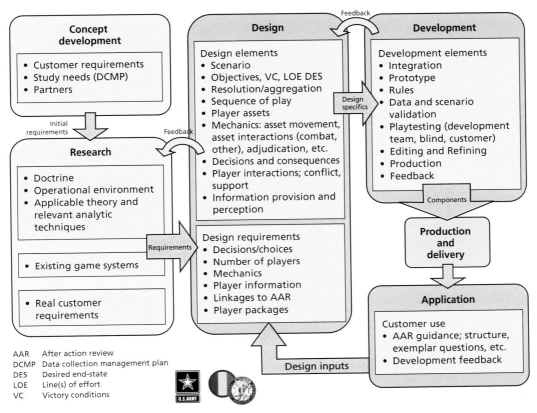

SOURCE: TRAC, "Wargaming Introduction and Applications," presentation for the RAND Corporation, Santa Monica, Calif., August 31, 2016b.

- be interesting and playable enough to make players want to suspend their inherent disbelief and open their minds to an active learning process
- be accurate and realistic enough to ensure that the learning that takes place is informative and not misleading.

A wargame should meet several criteria to be judged successful. The wargame should meet study requirements in a timely fashion, with results that are relevant and emerging concepts identified. Decisionmakers should understand the utility of the game, as well as its benefits and limitations, and wargame results should help inform decisions. Participants should be energized, feel a sense of camaraderie, and have fun. Additionally, the event should go smoothly, and lessons should be identified that can be applied to future events.

The University of Foreign Military and Cultural Studies

Overview

UFMCS at Fort Leavenworth views itself as the premier purveyor of critical thinking within the U.S. Army, and the core of its curriculum is based on applied critical thinking, fostering cultural empathy, self-awareness and reflection, and groupthink mitigation. UFMCS incorporates a number of methods, techniques, and frameworks from a variety of sources.[71] These include approaches found among the intelligence community's SATs.[72]

Referred to collectively as red teaming, the set of methods and approaches taught by UFMCS is meant to create leaders who can facilitate more divergent and creative thinking in groups to make better decisions. The aftermath of the 2003 Iraq invasion, in particular, prompted discussion within the Army about how to encourage more divergent ways of thinking. Outside the Army, the primary adopters of UFMCS's methods have been the Marine Corps, Special Operations Forces (SOF), Customs and Border Protection, and Strategic Command. Within the military services, the Marine Corps was an early adopter of red teaming. UFMCS has also supported red team training for the Department of State and the U.S. Agency for International Development.[73]

Wargaming and the University of Foreign Military and Cultural Studies

UFMCS does not wargame, but is included in this report because red teaming methods are used by some wargaming centers, such as the CSL. USMCS has also taught a shortened version of its red teaming class to wargamers at the 2016 and 2017 MORS special meetings on wargaming. MORS organizers introduced red teaming to the wargaming community in order to expand the knowledge of different structured group techniques available for times when adjudicated games are not necessary but greater structure is still desired. Participants in the class also concluded that applying red team methods to wargame processes and designs could improve wargames themselves.[74]

Tools and Approaches

UFMCS teaches several variants of its red teaming class. Its nine- and 18-week red teaming leader courses give soldiers an additional skill identifier. (The 18-week course is conducted almost exclusively for the SOF community.) UFMCS also has a six-week course that qualifies participants as red team members, as well as one- and two-week courses that can be offered by its mobile training teams.[75]

---

[71] U.S. Army Combined Arms Center, "University of Foreign Military and Cultural Studies/Red Teaming," webpage, undated b.

[72] U.S. Central Intelligence Agency, 2009.

[73] Interview with UFMCS staff, Fort Leavenworth, Kan., August 30, 2017. MCU is also an adopter of red teaming, and OAD has hosted red teaming classes for its analysts.

[74] Pournelle, 2017, p. 6.

[75] Interview with UFMCS staff, Fort Leavenworth, Kan., August 30, 2017.

Table A.9
University of Foreign Military and Cultural Studies Wargaming Tools and Approaches

Tool or Approach	Description	Usage
The Applied Critical Thinking Handbook	Handbook of structured group methods aimed at reducing bias, mitigating groupthink, and encouraging critical thinking.	Used for framing questions, generating approaches, and evaluating potential approaches.

UFMCS makes several handbooks available to those interested in red teaming. Among its primary tools, which are listed in Table A.9, is the *Applied Critical Thinking Handbook*, currently in its eighth edition.[76] Other red teaming tools that UFMCS makes available include the *Liberating Structures Handbook* and the *Group Think Mitigation Guide.*[77]

Facilities

UFMCS has classroom spaces, and also offers shortened versions of the red teaming course on-site for hosting organizations.

Process and Skill Sets

Red teaming encompasses a diverse set of group methods geared at enabling divergent thinking, challenging assumptions, and enabling alternatives. UFMCS considers a graduate of one of its red team leader courses to be capable of leading red teams.

Key Best Practices and Recommendations

To be successful, red teams require protection from superiors to preserve their unique function. As a rule of thumb, red team members should not be short-term participants but part of a process. Red teamers should have the ability to work on many issues and operate without a portfolio. However, red teaming should not be their permanent job. UFMCS estimates that red teamers should be time limited to 18 months.

U.S Air Force Wargaming Centers

As with the other services, we were unable to interview members of all the organizations within the Air Force that wargame. In particular, we note that we did not have enough time to reach out to the many excellent wargamers at Air Force Headquarters, Air Force Space Command, the Air Staff, and the other major commands.

---

[76] UFMCS, *The Applied Critical Thinking Handbook*, version 8.1, Fort Leavenworth, Kan.: UFMCS, September 2016.

[77] UFMCS, *Liberating Structures Handbook*, Fort Leavenworth, Kan.: UFMCS, undated a; UFMCS, *UFMCS Group Think Mitigation Guide*, Fort Leavenworth, Kan.: UFMCS, undated b.

Air Force Materiel Command
Overview
AFMC is responsible for the research, development, test, and evaluation process; acquisition management; and worldwide logistics support to Air Force weapon systems. Its mission covers activities ranging from maintenance, modification, and overhaul of weapons, to S&T, to running the Air Force's medical and test pilot schools. AFMC accounts for nearly one-third of the Air Force budget.[78] The AFRL, an organization within AFMC, has a separate entry in this report.

Wargaming at Air Force Materiel Command
Headquarters AFMC provides guidance for AFMC's role in the initiation, development, and execution of wargames. It is responsible for providing subject matter continuity for wargaming within the command and a single face to the wargaming community outside the command; leading all wargame integration activities for AFMC-level, Air Force Title 10, and other wargames; and collaborating with command directorates and centers on AFMC directorate- and center-level games.

The AFMC centerpiece for promoting wargaming continuity is the Wargame Working Group, which includes representatives from key command directorates and centers, as well as non-AFMC agencies. Members attend specialized wargame training and support regular Wargame Working Group meetings to assist in AFMC's wargame processes. Typically these meetings are conducted monthly to keep Wargame Working Group members informed on all AFMC wargaming activities and findings in order to influence the system, planning, and prioritization decisions in each command directorate and center.

AFMC is responsible for representing all Air Force materiel concepts for Headquarters Air Force–sponsored Title 10 and other Air Force–sanctioned wargames. Materiel concept integration must not diminish the realism or credibility of the wargame. While the AFRL leads all Air Force materiel concept development activities for systems, AFMC ensures proper representation of future materiel concepts and all aspects of the AFMC mission set during the wargame cycle (preparation, execution, feedback, and results). AFMC leads materiel concept calls for Air Force wargames, and collaborates with the AFRL on the analysis and vetting of all materiel and non-materiel concepts into wargames in support of Air Force games. AFMC also leads the down-selection and integration of all materiel and nonmateriel concepts into Air Force wargames.

AFMC often acts as wargame sponsor or lead. AFMC cosponsors and coleads the Global Mobility Agile Combat Support mini-wargame to ensure a realistic and credible Global Engagement capstone event representing air mobility and AFMC missions. Most significantly, it produces a Time Phased Force Deployment document for Global Engagement planners. It also prepares wargame analysts, adjudicators, and blue force and

---

[78] AFMC, "AFMC History" and "AFMC Mission," webpage, undated.

red force members. AFMC leads the Long Duration Logistics Wargame to analyze the Air Force's ability to logistically sustain combat operations in a contested scenario eight to 12 years in the future. Wargame personnel will support the Agile series of wargames, Weapons and Tactics Conferences, tabletop exercises, and other AFMC efforts.

AFMC and the AFRL conduct a future capabilities materiel concept down-selecting S&T wargame for materiel concept development. The AFRL conducts small materiel concept development games, called Future Analytical Science and Technology (FAST) wargames, with the goal to develop, validate, and verify materiel concepts for wargame inclusion.

Tools and Approaches

AFMC wargaming tools and approaches include spreadsheet tools, the Caffrey Triangle, and an AFMC-authored course on wargame development. The AFMC wargaming course is for official use only, and course material may be available to other U.S. government organizations. AFMC also uses the AFRL FAST games to refine adjudication in its games. Table A.10 summarizes AFMC's major tools and approaches.

Presently, AFMC finds it difficult to execute M&S for adjudication during the course of a game. It is thus exploring other tools to support decisionmaking during a wargame, including commercial software from Australia and the United Kingdom. This would help address questions about such issues as the placement of orbits. AFMC is aiming to acquire fairly simple software to accomplish such tasks in game time.

Figure A.5 illustrates the Caffrey Triangle, which serves as a framework for considering the purpose of a red team in a game. Additional information on its use can be found in Appendix B.

Facilities

AFMC does not have its own wargaming facilities. It may use facilities at the AFRL, the National Air and Space Intelligence Center, or the Pacific Warfighting Center.

Table A.10
Air Force Materiel Command Wargaming Tools and Approaches

Tool or Approach	Description	Usage
AFMC wargaming course	For-official-use-only wargaming course developed by AFMC to teach how to run a wargame.	Used by AFMC students to conduct wargames.
AFRL FAST games	Adjudication of future technology concepts against Title 10 scenario target sets.	FAST game results are used for other AFMC wargame adjudication.
Caffrey Triangle	Framework for considering the purpose of a red team in a game.	Used to assist discussion about the type of opponent appropriate for a wargame.
Spreadsheet tools	Excel spreadsheets and macros.	Used to assist in planning, adjudication, and analysis.

Figure A.5
The Caffrey Triangle

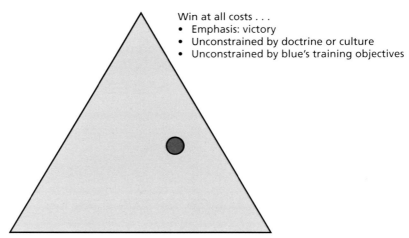

Win at all costs . . .
- Emphasis: victory
- Unconstrained by doctrine or culture
- Unconstrained by blue's training objectives

Train me . . .
- Emphasis: a foil for blue
- Unconstrained by doctrine or culture
- Constrained by blue's training objectives

Act like "them" . . .
- Emphasis: realism
- Constrained by doctrine or culture
- Unconstrained by blue's training objectives

SOURCE: J-8 SAGD.

Process and Skill Sets

Skilled personnel that AFMC employs in its wargames include OR analysts and various SMEs who understand Air Force operations.

AFMC has been historically associated with running the annual Connections Wargaming Conference, which serves as a focal point for many defense and commercial wargamers. The network of organizers and contributors forms a CoP that other defense wargamers are able to turn to for wargaming knowledge and methods.

AFMC also teaches wargame design, execution, and analysis internally and draws from these Air Force personnel as a way of developing staff with wargaming skills. Several versions of this course are available. Course guidelines are to develop and execute wargames according to the following series of steps:[79]

1. defining the task from the sponsor
2. designing the game (including an alpha test)
3. developing, researching, and drafting a wargame kit (including a beta test)
4. preparing administrative and logistical aspects of the game
5. rehearsing the game

---

[79] Matthew B. Caffrey, Jr., "Wargaming: Design & Development," briefing, Headquarters, Air Force Materiel Command, Wright-Patterson Air Force Base, Ohio, January 11, 2017; Matthew B. Caffrey, Jr., "Completing the Wargame Cycle: Event & Item Wargames," briefing, Headquarters, Air Force Materiel Command, Wright-Patterson Air Force Base, Ohio, January 11, 2017.

6. executing the game
7. conducting wargame analysis
8. writing the wargame report
9. distributing and archiving the wargame report.

Key Best Practices and Recommendations

AFMC focuses on getting accurate information on concepts and to get high fidelity games. Games at AFMC attempt to follow five principles:

1. concept descriptions have to be accurate
2. blue players must employ concepts and systems accurately
3. adjudication must be correct
4. red players have to counter blue players appropriately
5. people with experience in analysis have to assess the game.

The Air Force Research Laboratory

Overview

Based at Wright-Patterson Air Force Base, Ohio, the AFRL is tasked with leading the discovery, development, and integration of affordable warfighting technologies for American air, space, and cyberspace operations. Organizations within the AFRL address aerospace systems, directed energy, information, munitions, materials and manufacturing, sensors, space vehicles, technology transfer, and human performance.[80] The AFRL falls under AFMC.

Wargaming at the Air Force Research Laboratory

The AFRL's collaboration with AFMC on wargaming is described in the section above on AFMC. In response to Deputy Secretary of Defense Robert Work's 2015 memorandum on wargaming, the AFRL has "taken steps to revitalize wargaming across its Enterprise." Steps taken to better manage wargaming include establishing a Wargaming Working Group, creating a vetted concept portfolio, forming a cadre of wargaming technologists, and promoting M&S for wargaming support.[81] The AFRL also held a workshop in May 2016 to establish baseline M&S tools to support wargaming. It began its FAST game series in August 2016, focusing on advanced technology concepts in Anti-Access/Area Denial environments.[82]

Figure A.6 shows the AFRL's analytical wargaming framework, and lays out tasks and activities associated with wargaming training, design, execution, and reporting:

Organizationally, the top priority for AFRL wargaming is to support topics of interest to the commanding general of the AFRL. The laboratory also supports Air

---

[80] Wright-Patterson Air Force Base, "Air Force Research Laboratory," webpage, undated a.

[81] Bestard, 2016, p. 12.

[82] Bestard, 2016, p. 13.

Figure A.6
The Air Force Research Laboratory Analytical Wargaming Framework

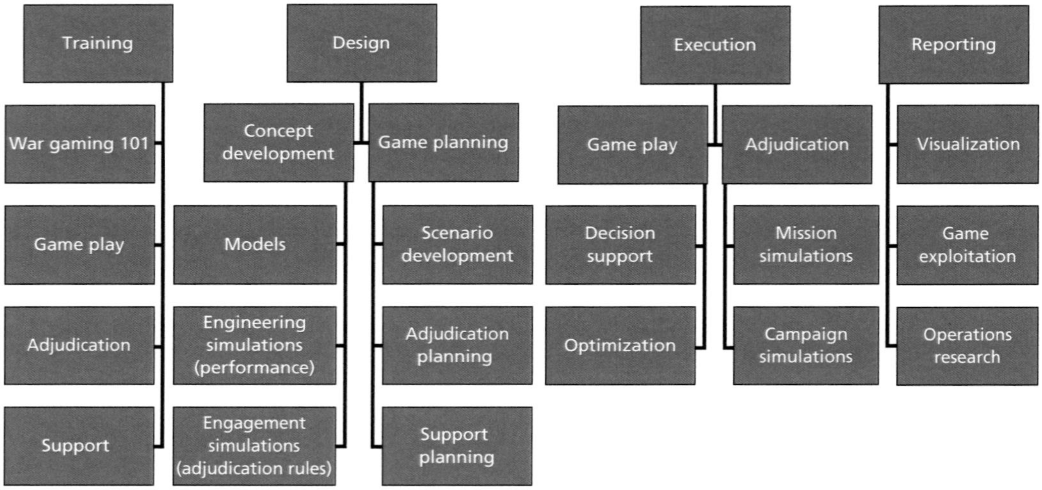

Training	Design		Execution		Reporting
War gaming 101	Concept development	Game planning	Game play	Adjudication	Visualization
Game play	Models	Scenario development	Decision support	Mission simulations	Game exploitation
Adjudication	Engineering simulations (performance)	Adjudication planning	Optimization	Campaign simulations	Operations research
Support	Engagement simulations (adjudication rules)	Support planning			

SOURCE: Jamie J. Bestard, "Air Force Research Laboratory Innovation: Pushing the Envelope in Analytical Wargaming," special issue, "Modeling and Simulation Special Edition: Wargaming," *CSIAC Journal*, Vol. 4, No. 3, November 2016, Figure 1.

Force Headquarters in its Futures Game and Global Engagement, where it works to ensure that proposed concepts are represented accurately. The AFRL also provides technical support to Unified Engagement, the Air Force's Title 10 series of wargames. Other AFRL wargaming includes support to the Air Force major commands; the FAST game series; and smaller, quick-turn wargames. AFRL wargamers also find themselves putting together "quivers" of concepts that are actually being worked on in the labs, and representing these concepts in M&S.[83] Wargaming at the AFRL appears to fall largely within wargaming for concept development and S&T wargaming.

A key attribute of AFRL gaming is the prevalence of M&S before, during, and after wargames. Due to the importance of technical details in the projection of air power on the battlefield, the AFRL emphasizes the accuracy of technology and realistic adjudication within its wargames. It supports wargames with external sponsors, but has recently begun to sponsor its own wargames. In the future, the AFRL aims to utilize uniform service members in developing CONOPs that can be leveraged in wargames.[84]

Tools and Approaches

A summary of some key tools and approaches that we discussed with AFRL are presented in Table A.11. JSAF was previously used by the AFRL Human Effectiveness

---

[83] Telephone interview with wargamers at AFMC and AFRL, October 4, 2016.

[84] Telephone interview with wargamers at AFMC and AFRL, October 4, 2016.

Table A.11
Air Force Research Laboratory Wargaming Tools

Tool	Description	Usage
FAST	Series of vignettes, based on Title 10 game scenarios and target sets, that enable players to utilize a wide range of technology in the proof-of-concept phase of development to better understand potential impact on the battlefield.	Concept refinement: to test whether and how new technology can be useful in solving tactical and operational problems, and particularly qualitative output. FAST allows for quick examinations of plausible technology, before serious study of feasibility and broader buy-in occurs as part of the Title 10 process.
Panther Games artificial intelligence engine	Commercially developed artificial intelligence engine for command and control simulations.	
Unified Engagement technical support	AFRL provides support representing emerging technology and capabilities into the Title 10 Unified Engagement game. These tools are largely Excel macros.	Adjudication and postgame analysis.

Directorate; more about JSAF can be found in Appendix B. However, we do not include it in the table, as it did not come up as a key, current tool used in AFRL wargaming. The AFRL has also previously been a sponsor of the Green Country Model (GCM), which can also be found in Appendix B, but we do not list the GCM here for the same reason.

The AFRL is also looking into a number of other software tools that it hopes it can apply to wargaming. This includes an artificial intelligence engine for command and control commercially developed by an Australian company, Panther Games; commercial PC games developed by a UK company, Slitherine; and Warlock, developmental software that will help integrate information during games. The AFRL is also looking for tools to support decisionmaking during a wargame.

Each technology directorate within the AFRL tends to own its unique set of software. While the AFRL writ large uses a number of M&S tools at the engineering, engagement, mission, and campaign levels, M&S tools are often not at the point of development at which they can be used in real time during a wargame.

Facilities

The AFRL does not have dedicated wargaming facilities, but instead relies on the Air Gaming Center.

Process and Skill Sets

The AFRL's wargaming staff comprises six members, two of whom are M&S focused. Like many other service gaming centers, the AFRL leverages participation by other DoD personnel to provide expertise that may be relevant for a given game. The AFRL draws on such organizations as the Air Force Institute of Technology and the National

Air and Space Intelligence Center. Support staff at the Air Gaming Center also provide administrative support.

In terms of processes, a game in the FAST series takes five people six weeks to prepare. The team takes a part of a Title 10 scenario and one of its target sets, setting aside ground and naval combat. FAST games use three different adjudication methods, two of which are computerized. Due to the AFRL's focus on new technologies, it utilizes FAST wargames to prepare realistic parameters for larger wargames as a form of concept refinement. FAST is not intended to validate concept performance but to serve as an experimental test of a concept's potential applicability to various tactical and operational scenarios.

Key Best Practices and Recommendations

To ensure feasible technologies and operational expectations, AFRL has prioritized accuracy of acquisition and fielding of technology within its wargames. Planners recognize acquisition matters more on Title 10 games and whether an industrial base exists for development. At the moment the AFRL is attempting to figure out how much life cycle management needs to be accurately represented.

Due to the high level of technical sophistication within AFRL wargames, its wargaming program has developed a series approach where its FAST wargames inform its larger Title 10 wargames. This method may prove informative for other service wargaming centers in their own Title 10 wargames. Presently the AFRL is prioritizing archiving and documenting the ideas that come out of wargaming so that when a discussion reoccurs in later years, results can be referenced.

The Air Universitys LeMay Center for Doctrine Development and Education
Overview

The Air University's LeMay Center for Doctrine Development and Education identifies its mission as developing warfighters through doctrine, education, and wargaming.[85] Responsibilities related to doctrine include developing and revising service basic and operational-level doctrine, developing the Air Force position in joint doctrine, and contributing to North Atlantic Treaty Organization (NATO) and allied doctrine. The LeMay Center also plays a substantive role in senior leader education, professional continuing education, and activities to teach doctrine through online and resident courses. Wargaming is also a part of its portfolio.[86]

Wargaming at the LeMay Center

Gaming at the LeMay Center largely involves wargaming to support joint and service education, wargaming to support concept development, and wargaming to support operational decisions and plans. In terms of size, the largest wargames can involve up

---

[85] Air University, "Curtis E. LeMay Center for Doctrine Development and Education," webpage, undated.

[86] Air University, undated.

to 500 ACSC students who need a COP across air, sea, space, and ground forces; information about a large order of battle; a future database of forces; and a trained team of controllers. On the other end of the scale, wargames to support concept development tend to be small, with less than 15 people examining a concept over one or two days. On average, LeMay has hosted up to 50 wargames a year.[87]

JLASS, an education wargame involving the senior military war colleges, is LeMay's largest in-house effort, with approximately 200 participants.[88] As a theater-level wargame, JLASS stresses joint and combined warfare where participants must analyze various U.S. military responses to regional crises. JLASS strives to enhance JPME.[89] Preparing for JLASS is typically a year-round effort, with team members traveling once quarterly to meetings with all the colleges. The various colleges develop JLASS scenarios over the summer and fall; the actual JLASS wargame takes place over five days in the spring. Though this event has reached a steady state of expertise over twenty years, there are constant modifications to enhance the quality of gameplay.[90]

The LeMay Center has conducted a number of other games to support education:[91]

- Global Challenge, a capstone wargame for the students of the Air War College
- Joint Air Exercise, the ACSC Joint Air and Space Operations course wargame that covers air and space operations
- the Joint Intermediate Planning Staff Exercise, a collaborative planning joint task force exercise with ACSC and the CGSC
- the Joint Planning Exercise, the culminating event for the ACSC Joint Campaign Planning course
- Joint Wargame, the culminating event for the ACSC's Joint Warfare Studies course
- Judge Advocate Wargames, a variety of capstone events for students of the Air Force Judge Advocate General School to test their communication, management, and leadership skills in an operational environment
- Pegasus, a computer-adjudicated wargame between two opposing Combined Forces Command staffs in a notional conflict between two fictional alliances.

Aside from educational wargames, the LeMay Center supports the Futures Game and Global Engagement, which are Chief of Staff of the Air Force Title 10 games. The Futures Game compares alternative futures and force structures, and is an input

---

87 Telephone conversation with LeMay Center personnel, April 4, 2017.

88 Telephone conversation with LeMay Center personnel, April 4, 2017.

89 Air University, "U.S. Air Force Wargaming Gateway," webpage, March 12, 2019.

90 Telephone conversation with LeMay Center personnel, April 4, 2017.

91 Air University, 2019.

into the Air Force's decisions on future concepts and force structure.[92] It also uses other operational games to support concept development and refinement. For example, LeMay supports the Schriever Wargame to explore critical space issues, concept development games for small tankers, medical logistics, cyberoperations, and future weapons concepts. Frameworks and COAs are particularly interesting to LeMay.[93]

The LeMay Center also supports wargames for operational decisions and plans. This includes a number of games for commands and Air Force agencies such as the AFRL, Air Force Global Strike Command, the 18th Air Force, and Pacific Air Forces.[94]

Tools and Approaches

LeMay utilizes various tools and approaches in the design and execution of its wargames. One key tool is LeMay's Wargaming Gateway, a content management system developed in-house to support wargames. Elements in this content management system include a request for information system, embedded email, a document manager, and clock mechanisms, which trigger event injects in a game. It also includes a survey employing a Lickert scale that is able to do system tallies for reports. Additionally, the Wargaming Gateway includes a COP web application, Web Forces Online, which displays the world and different air, land, and maritime orders of battle. Students can access the application on the internet and familiarize themselves with the material throughout the academic year.

Other tools that Air University colleges have used include spreadsheet tools and commercial PC games. The Squadron Officer College and the ACSC use PC-based games as real-time strategy exercises, which are adjudicated live. Examples of commercial PC games they use include *JAEX* and *Modern Air Power* from John Tiller Software; in the past, schools at the Air University have also used *StarCraft*, *Star Wars Battlefront*, and *Rome: Total War*. Table A.12 shows the major tools that LeMay uses for wargame support. The center also has instances of games that only use paper maps or no maps. People we spoke with at LeMay were able to cite one game that used only AV equipment.

Facilities

LeMay is currently upgrading the security level of its facility, which has 25 game rooms, large conference rooms, and office space for staff. LeMay facilities also host wargames for other Air Force organizations.

Process and Skill Sets

The skill sets that LeMay finds useful to support wargames include designing scenarios and visuals to support games, software development, and analysis. In the past, LeMay

---

[92] Air University, 2019.

[93] Telephone conversation with LeMay Center personnel, April 4, 2017.

[94] Air University, 2019.

Table A.12
LeMay Center Wargaming Tools and Approaches

Tool or Approach	Description	Usage
Commercial PC games	Commercially available computer games.	Used in real-time strategy exercises for students.
Wargaming gateway	Internally developed content management system with a variety of wargame support tools.	Used for gameplay.
Spreadsheet tools	Excel-based tools.	Used for accounting, adjudication, and data capture.

retained roughly 40 software developers, but currently the staffing is at about 15 and is a mixture of both military and civilian developers. Analysis is another area where LeMay has experimented with other personnel, such as social scientists. LeMay would like to have dedicated analysts to review the full commentary from games, but does not have that level of staffing.

The process for LeMay's largest game, JLASS, typically lasts a year. Preparations for the game begin with scenario development among the schools over the summer, a distributed period in the fall (often October), when staff begin to develop the game with software tools and begin to make information available to students online, and in-person play the following spring (typically March).

The LeMay Center also described its process for an example concept game. The first step is to identify the concept. The second step, which the center views as the most important, is to identify the game objective. The game objective will drive the scenario, the adjudication experts, the players, and the control team needed, as well as where simulation tools may be appropriate. Concept games typically require two to three months from inception to actual game, not including analysis. This time frame is for games where there are already existing concepts and no new simulation tools need to be created. Concept games often use a "shoot—look—shoot" scenario over the course of a two-day game. LeMay staff then creates a summary report within one to two weeks. The entire process for a concept game may thus take 90 days rather than a year.

In terms of game analysis, after action reviews are much simpler for education games than for concept development. By the end of an educational game, the faculty generally knows whether it has achieved the educational objectives. It will also have student survey results from the game to give additional information. For concept games, LeMay captures data and comments, and has experimented with using social scientists to look at decisionmaking. While LeMay captures comments with data tags using Excel tools, it does not have the full set of analysts it desires to analyze all the data it captures.

Key Best Practices and Recommendations

LeMay argues that educational games should provide decisionmaking experiences, whereas concept development games provide decisionmaking information, which leads to more research. In concept development, the experience of the player is less of a focus, and the main concern is the accuracy of the results and the decisions the players made.

Office of the Secretary of Defense Wargaming Centers

Three OSD offices are chairs in DWAG: OSD CAPE, OSD ONA, and OSD Policy. However, ONA and OSD Policy largely outsource their wargames and do not maintain an internal team of wargamers. The one exception is CAPE.

Cost Assessment and Program Evaluation

Overview

As part of the OSD, CAPE provides DoD with independent and critical analysis on resource allocation and cost estimation problems to deliver the optimum portfolio of military capabilities.[95] CAPE is responsible for the management of DoD's programming process, including the development of planning guidance (in conjunction with other OSD organizations), production of applicable programming guidance, and direction of the annual program review.[96]

Wargaming at Cost Assessment and Program Evaluation

Wargaming within CAPE is overseen by the Strategic Analysis and Wargaming Division; the Simulation and Analysis Center is the execution arm. CAPE utilizes analytical wargames, which are incorporated through a wider long-term cycle of research. Ultimately, CAPE wargames are designed to inform the PPBES within DoD. CAPE wargames have also been used to support program review, OPLAN development, nuclear posture review, and the National Strategy Review.[97] Overall, we consider CAPE wargames to largely function as support to capability development and analysis.

CAPE specializes in analytical wargames aligned with its program management mission. According to CAPE, an analytical wargame can best be characterized as "the act of competitive, contextualized decision-making within pre-defined constrains for the purpose of gaining insight into complex, adaptive, interactive, and cognitive systems."[98] CAPE prefers long-term and multimethod approaches over isolated, single-use wargames, and believes it must be able to have analytic rigor and defensible insights from wargames.

Within its analytical wargames, CAPE strives to understand what is successful and to set up a process where ideas, biases, and perspectives compete. The purpose of

---

[95] OSD CAPE, "About CAPE," undated.

[96] OSD CAPE, undated.

[97] OSD CAPE, "Wargaming in CAPE," briefing at Connections 2017, Quantico, Va., June 1, 2017.

[98] OSD CAPE, 2017.

this competition is to generate a "theory of success" that is then subjected to falsification tests using other analytical means. As a result, analytical wargaming requires two critical components for success: successive learning over multiple iterations of the wargame and the validation of underlying assumptions through other means within Peter Perla's cycle of research.[99]

One way that CAPE may differ from other wargaming organizations is that while other organizations may often use outside experts in their wargames to generate insights, CAPE's approach is to develop internal expertise through a continuing process of gameplay and analysis. CAPE finds that this process of building internal learning and expertise, where its staff are the players, allows it to leverage this knowledge in subsequent studies and wargames.[100]

CAPE conducts numerous games every year. For example, there may be 20 games related to a single study. The vast majority of CAPE games are rigidly adjudicated, and the rigid adjudication game systems are often based on commercial wargame designs that have heavily modified rule sets. CAPE looks for commercial board wargames that are designed at a similar scale and subject and mine those systems for useful models. It also develops its own rule sets for system interactions for which there are no commercial gaming analogs.[101] Games also range from the tactical level (with highly rigid adjudication), to the operational level (also with rigid adjudication), to pol-mil games at the strategic level.[102]

CAPE also conducts a small number of games that are not rigidly assessed, such as capstone events. Game adjudication in these cases can involve self-adjudication or SME adjudication.[103]

Tools and Approaches

CAPE conducts most of its games using physical maps and playing pieces.[104] Its major wargaming tools are modified commercial board games, spreadsheet tools for adjudication, R, and SWIFT, whose development was sponsored by CAPE.[105] These tools are listed in Table A.13. Other tools that CAPE uses in wargaming include network modeling tools, MATLAB, and existing systems models.[106]

SWIFT, a tool developed by CAPE, can be useful in designing wargames with a strong visualization aspect. It can also be useful for recording game moves, collecting

---

[99] OSD CAPE, 2017; Perla, 2012, pp. 251–253.

[100] Email from OSD CAPE staff, September 19, 2017.

[101] Email from OSD CAPE staff, September 19, 2017.

[102] Interview with OSD CAPE staff, Arlington, Va., October 17, 2016.

[103] Interview with OSD CAPE staff, Arlington, Va., October 17, 2016.

[104] Email from OSD CAPE staff, September 19, 2017.

[105] Interview with OSD CAPE staff, Arlington, Va., October 17, 2016.

[106] Email from OSD CAPE staff, September 19, 2017.

Table A.13
Cost Assessment and Program Evaluation Wargaming Tools and Approaches

Tool or Approach	Description	Usage
Modified commercial board games	Commercially available games for educational or analytical purposes.	Used for a variety of wargame topics, from COIN to conventional conflict.
R (Statistical Programming Language)	Statistical programming language used for statistical data analysis and visualization.	Used for wargame analysis.
Spreadsheet tools	Excel-based tools that run in the background during games.	Used for adjudication.
SWIFT	A platform for capturing gameplay, adjudication, data analysis, and constructing COP.	Used for bilateral and multilateral wargames, logistics, personnel recovery, cyberoperations, and exploring future conflicts with Russia.

data, adjudication, visualization, and presentation. SWIFT has been used on unclassified networks, the Joint Worldwide Intelligence Communications System, and SIPRNet. However, depending on the degree of customization required, one full-time programmer may be needed for SWIFT support when developing games.[107]

Facilities

CAPE does not have dedicated wargaming facilities. However, it does have access to multipurpose spaces, including large spaces and multiple rooms at the government facility where CAPE is based.

Process and Skill Sets

CAPE wargames strive to demonstrate and record the evolution of learning through multiple wargame iterations given the constraints of the wargame's objectives. This is achieved through an extensive process of research, literature review, wargaming, and analysis. For a game, CAPE reviews what the customer wants, identifies participants, adjudicates the game, collects data, and conducts ex post facto analysis.[108]

The CAPE Simulation and Analysis Center has 10–12 staff members who may be involved in putting on a wargame, and may also borrow a half dozen others. CAPE looks for a variety of quantitative skills in its wargaming staff, such as abilities in math, statistics, data analysis, formal logic, econometrics, M&S, OR, and game theory. At the same time, CAPE seeks individuals with backgrounds in international relations, political science, economics, history, military history, sociology, anthropology, business, education, and psychology. Other practical skills that it seeks in a wargame team

---

[107] Interview with OSD CAPE staff, Arlington, Va., October 17, 2016.

[108] Interview with OSD CAPE staff, Arlington, Va., October 17, 2016.

include art and graphic design, facilitation, writing, research, game design, and hobby gaming.[109]

One important element in CAPE wargaming that is relevant for processes and skill sets is its use of wargaming to develop internal expertise. While wargames traditionally draw from outside experts to help generate insights, CAPE uses a continuing process of gameplay and analysis to develop its own experts. This allows CAPE to leverage this internal learning and expertise in future studies and games. CAPE brings in outside participants more as reviewers than as players. The ultimate purpose is to create a deep body of research within CAPE on complex problem sets.[110]

Key Best Practices and Recommendations

CAPE has refined its analytical wargaming for the sole purpose of informing its programmatic mission within the OSD and DoD. Although the breadth of its wargaming may be narrower than other wargaming centers, CAPE emphasizes incorporating rigor and diversity within its analysis and wargaming. CAPE's long-term and reiterative method of wargaming is reflective of a broader trend within the DoD enterprise of embracing cumulative series of wargames rather than isolated, single-use wargames.

Another practice that CAPE emphasizes is the process of fitting the game to the question being analyzed. It is important to have game designs that are specific to the problem rather than reusing designs and possibly applying them inappropriately.[111]

Joint Staff Wargaming Centers
The J-8 Studies, Analysis, and Gaming Division
Overview

Established by President Dwight D. Eisenhower in 1947, SAGD was originally called the Army Advanced Study Branch, sought to evolve concepts relevant to national security. Throughout the years, the organization evolved and assumed various missions, including the establishment of the J-8 and the Political-Military Assessment Division within the Joint Staff in 1986.

J-8 SAGD is a strategic wargame and assessments office supporting the Chairman of the Joint Chiefs of Staff and DoD. Largely serving senior leadership from the Joint Staff and the CCMD, J-8 conducts joint, bilateral, and multilateral wargames and interagency politicomilitary seminars and simulations.[112] J-8 wargames include the CJCS Strategic Seminar Series, Interagency Policy Committee games, and multilateral games with international partners. On average, J-8 executes roughly

---

[109] Interview with OSD CAPE staff, Arlington, Va., October 17, 2016.

[110] Email from OSD CAPE staff, September 19, 2017.

[111] Email from OSD CAPE staff, September 19, 2017.

[112] Margaret McCown, "Defense Wargaming Panel: The SAGD/Provider Perspective," presentation at the Connections Wargaming Conference, Quantico, Va., August 2017; Joint Staff, "J-8 Force Structure, Resource & Assessment," undated.

two to three tabletop exercises per month, but the operational tempo can fluctuate to accommodate the needs of senior leadership. From January 2014 to November 2016, J-8 conducted roughly 80 wargames and decision support games.[113]

Wargaming at J-8 Studies, Analysis, and Gaming Division

J-8 mainly specializes in seminar-style tabletop exercises aimed at senior leadership, which we categorize for the purposes of this study as senior leader engagement and strategic discussion. In terms of topics, J-8 wargames focus on high-level strategic, policy, capabilities, and pol-mil wargames. Unlike service-oriented wargaming programs, J-8 wargames typically address how to frame a particular problem set or strategic challenge rather than focusing on testing or producing a specific solution. One of the key attributes of J-8's wargaming program is its access to senior leadership across the national security enterprise and its dedicated secure wargaming facilities. The participants and audience for SAGD games are typically senior leadership. Due to the time constraints that most senior leaders face, most of J-8's wargames usually last only a few hours and no longer than a day. Some of its wargames do incorporate red teaming, depending on the scenario. Overall, J-8 SAGD aims to develop insights in ill-structured problems, posit alternative thinking, and generate novel approaches in its games.[114]

Tools and Approaches

Due to the nature of its wargames, J-8 does not utilize particularly complicated tools or time-consuming approaches. Instead, most of its wargames rely on the experience of its senior players and the expertise of its staff. J-8 wargames are aimed at generating ideas and eliciting information from stakeholders. However, J-8 does employ tools like Decision Lens and ThinkTank to encourage discussion. J-8 also uses the Caffrey Triangle during discussions with sponsors to get a sense of how they may prefer adversaries to be represented in a game.[115] A summary of the tools and approaches utilized by J-8 is presented in Table A.14.

Facilities

One of reasons that J-8 continues to draw sponsors is its dedicated, secure wargaming facility in the Pentagon. J-8 is able to provide the facilities and staff support required for Top Secret SAP/SCI games. J-8 believes that this capability, combined with VTC capabilities, allows it to monopolize the market on high-level policy and strategy wargames run by the government in the Washington, D.C., metropolitan area. However, a drawback of a secure facility is the inability to quickly adopt commercially available emerging technology due to security restrictions.[116]

---

[113] McCown, 2017.

[114] Interview with J-8 SAGD staff, Arlington, Va., November 21, 2016.

[115] Interview with J-8 SAGD staff, Arlington, Va., November 21, 2016.

[116] Interview with J-8 SAGD staff, Arlington, Va., November 21, 2016.

Table A.14
J-8 Studies, Analysis, and Gaming Division Wargaming Tools and Approaches

Tool or Approach	Description	Usage
Caffrey Triangle	Framework for considering the purpose of a red cell in a game.	Used during discussions with wargame sponsors to discuss how to handle the adversary in a game.
Decision Lens	An end-to-end software solution and process for identifying, prioritizing, analyzing, and measuring which investments, projects, or resources will deliver the highest returns; and allows organizations to immediately see the impact and trade-offs of the choices they make.	Used to reduce risk and improve outcomes by facilitating collaboration. It allows stakeholders to more easily discuss key trade-offs and allocation decisions through scenario-based planning.
ThinkTank	Group decision support software for brainstorming, innovation, decisionmaking, and virtual interactive meetings.	Helps game participants have better, richer discussions. Groups can contribute anonymously, and then collectively, to vet all possibilities. Teams can then evaluate and prioritize the results of their input.

Process and Skill Sets

Each wargame is assigned a principal game director and a deputy game director, whose responsibilities begin with the inception of the wargame and carry through to its execution. During the execution phrase, the team will assume additional staff. From planning to execution, J-8 wargames routinely take roughly nine weeks. A skilled facilitator is crucial to successful game execution.

At that time of our interviews, SAGD wargaming staff consisted of eight military service members, five civilians, and three contractors. Of the roughly 16 staff members, expertise and backgrounds included foreign affairs personnel, strategic pol-mil specialists, active duty service members with operational experience, and OR analysts. Typical of small wargaming programs, J-8 is a flat organization whose staff assumes multiple roles depending on operational needs.[117]

Key Best Practices and Recommendations

One attribute of J-8 wargaming is its focus on high-level strategic and multiagency wargames. J-8 approaches wargaming from a whole-of-government mentality.

One recommendation from J-8 was not only to build to the user of tomorrow but to also take into consideration the user of today. For example, not all wargame participants are familiar or comfortable with highly advanced technical tools. J-8 also recommended that the Marine Corps send its core team of wargamers to work with and familiarize themselves with other wargaming programs around the country. By doing so, J-8 believes that the Marine Corps will be able to adopt and incorporate the

---

117 McCown, 2017.

best aspects of each program, build institutional relationships, and better add value to the Marine Corps' institutional knowledge.[118]

National Defense University, Center for Applied Strategic Learning

Overview

NDU in Washington, D.C., educates and develops national security leaders who are prepared to think "critically, strategically, and creatively." NDU encompasses a number of academic programs, including the College of Information and Cyberspace, the College of International Security Affairs, the Dwight D. Eisenhower School for National Security and Resource Strategy, the Institute for National Strategic Studies, the Joint Forces Staff College, and the National War College. NDU also conducts flag officer and senior enlisted education and falls under the Joint Staff.[119]

Wargaming at Center for Applied Strategic Learning

Center for Applied Strategic Learning (CASL) is the wargaming center that support NDU's component colleges. CASL also supports select other DoD sponsors. The center also runs wargames on behalf of OSD Policy under the National Security Policy Analysis Forum research program.[120]

CASL runs a diverse range of strategic and high-level operational games that support educational programs on various national strategy and operational planning topics. CASL wargames include a diverse range of strategic and operational planning topics. We categorize these games as wargames to support training and education and wargames to support senior leader engagement and strategic discussion.

CASL also "advises curriculum developers and faculty game directors on wargaming capabilities and techniques, while providing exercise/simulation expertise to the University curricula writers for course materials supporting joint operational planning simulations and the wargame portions of the JPME curricula."[121] Our interview at CASL confirmed that it is engaged in game support at other academic institutions and support lectures on game design.[122]

Tools and Approaches

CASL uses tools and approaches that encourage or facilitate strategic and operational discussions and insights. The center specializes in seminar-style exercises, but also uses computer, board, and card games to support its work. At the time of our interview, CASL was in the process of developing a matrix game about Libya but had not yet used it in the classroom. CASL did report using commercially available games

---

[118] Interview with J-8 SAGD staff, Arlington, Va., November 21, 2016.

[119] NDU, *Annual Report for Academic Year 2016 (AY16)*, Washington, D.C.: NDU, 2016, pp. 2, 5.

[120] CASL, "CASL History," webpage, undated.

[121] CASL, undated.

[122] Interview with CASL staff, August 26, 2016.

such as *Aftershock* with NDU students. CASL staff were also considering a stand-alone gaming network apart from NDU's commercial networks, NIPRNet, and SIPRNet.[123] The main types of wargaming approaches and tools utilized by CASL are summarized in Table A.15.

Facilities

As of 2013, CASL and the Joint Forces Staff College Wargaming Center were merged together in hopes of integrating the NDU enterprise and strategic and operational endeavors. Thus, CASL is comprised of two divisions, one at the Joint Forces Staff College campus in Norfolk, Virginia, and one at Fort McNair in Washington, D.C.[124] The Washington campus does not have dedicated wargaming spaces and uses classrooms and other multipurpose rooms. The Norfolk campus has a dedicated wargaming center with 20 distinct rooms that can be configured in multiple ways.

Table A.15
Center for Applied Strategic Learning Wargaming Tools and Approaches

Tool or Approach	Description	Usage
Board and card games	Games intended to introduce abstract concepts like deterrence and cooperation. More complicated games have been used to communicate strategic and operational concepts.	CASL uses a COIN-based board game for educational purposes.
Modified commercial board games	Use of commercially available games for educational or analytical purposes. For all but the simplest games, modification is generally needed to make rule sets more accessible and create a playable game to address the educational requirement.	Used for education and experiential learning on various topics, including international humanitarian relief.
National Security Policy Analysis Forum Seminar Game	Single-cell, three-move seminar game. In each move a scenario is presented, with a list of discussion questions for participants. There is no adjudication between moves.	Used for interagency games or other areas where information sharing and relationship building is important to future problem solving.
Professional facilitation training	Facilitation training designed for consultants working with clients to move through a strategic planning process.	Used for formal facilitation training to wargame designers and facilitators.
Videos/video studio	Ranging from a studio for live video production to camcorders and computers with editing software; several centers stress the utility of being able to produce "new" video to deliver game scenario briefings and injects for heightened impact.	Used for introduction briefings of the game scenario and scenario updates.

---

[123] Interview with CASL staff, August 26, 2016.

[124] CASL, undated.

Process and Skill Sets

The center consists of about twenty civilian, military, and contractor analysts. Organizationally, CASL maintains a mixture of civilian, contractor, and military expertise. At times CASL seeks various SMEs for specific wargames and projects from across the NDU and the larger defense enterprise. NDU CASL was also the first wargaming center that we know of that provided formal facilitation training for its wargame facilitators, which we consider an important best practice. In terms of facilitation training, CASL engages the services of a firm called Leadership Strategies.[125]

Key Best Practices and Recommendations

Best practices at CASL include formal facilitation training for its wargamers, the use of modified commercial and customized board and card games, and the use of video software to create videos for wargames. CASL was also considering actions that other wargaming centers have recommended, such as matrix gaming and a standalone network. CASL is also a very active participant in the wargaming CoPs, with CASL staff acting as longtime organizers for the Connections Wargaming series of conferences.

Combatant Command Wargaming Centers

U.S. Special Operations Command Wargame Center

Overview

With its headquarters at MacDill Air Force Base near Tampa, Florida, SOCOM coordinates special operations planning and provides SOF to U.S. missions worldwide. SOCOM missions include counterterrorism, COIN, civil affairs, foreign humanitarian assistance, foreign internal defense, hostage rescue and recovery, military information support operations, security force assistance, and special reconnaissance.[126] SOCOM is organized into service components and theater special operations commands.[127]

The SOCOM Wargame Center falls under SOCOM Headquarters and reports directly to the chief of staff at SOCOM.[128] The Wargame Center, in collaboration with stakeholders, designs and conducts Senior Leader Seminars and other events that allow senior leaders to understand complex problems and develop solutions. The intent is to create an environment that allows senior leaders to think together, develop a shared understanding of the problem and the way ahead, and make decisions.[129]

---

[125] Interview with CASL staff, August 26, 2016.

[126] SOCOM, "Headquarters USSOCOM," webpage, undated a.

[127] SOCOM, "USSOCOM Theater Special Operations Commands," accessed through the SOCOM homepage, undated b.

[128] Email from SOCOM Wargame Center staff, October 5, 2017.

[129] Telephone interview with SOCOM Wargame Center staff, November 2, 2016.

Tools and Approaches

The SOCOM Wargame Center utilizes design thinking with emphasis on several different approaches, including qualitative research. These approaches and other tools are summarized in Table A.16. Since Wargame Center games focus on senior leader discourse, with an emphasis on developing a shared understanding, the majority of the tools and approaches emphasize learning, designing, and prototyping possible

Table A.16
U.S. Special Operations Command Wargaming Tools and Approaches

Tool or Approach	Description	Usage
Gaming auditorium	Large auditorium style gaming facility consisting of a large map surrounded by auditorium style seating. Supported by extensive AV equipment to allow for presentation and VTC connectivity.	Used for senior leader seminar and map exercise style games.
Soft systems methodology (SSM)	Method of inquiry, learning and analysis, often used in business, when there are competing views on the definition of the problem.	Provides a structured way to bring SMEs together to think systematically about a range of problems by facilitating a venue where a range of interpretations can be identified, their assumptions made explicit, and various trade-offs examined.
Active Advantage	Software tool that allows the visualization of large and complex data sets, which can be transposed to the large Wargame Center map.	Used as a visual planning workspace to stream planner products as a visualization of a storyboard.
The Applied Critical Thinking Handbook	Handbook of structured group methods aimed at reducing bias, mitigating groupthink, and encouraging critical thinking.	Used for framing questions, generating approaches, and evaluating potential approaches.
ArcGIS	Application that offers a unique set of capabilities for applying location-based analysis to business practices. It enables greater insights using contextual tools to analyze and visualize data and features collaboration and sharing capabilities.	Creates deeper understanding by allowing analysts to quickly see where things are happening and how information is connected.
Design thinking	Human-centered methodology that focuses on creating empathy for the problem and producing prototype solutions geared toward innovation.[a]	Used as the methodology for game design. Allows for facilitators to craft techniques and a process to illicit novel and useful ideas.
Systems thinking	Systems to "analyze interactions between various factors affecting a situation, to understand cycles of influence that affect our ability to intervene in changing the situation."[b]	Used to dissect a complex problem into manageable visuals to see how actors interact with each other, what the strengths and weaknesses are in the system, and how an effects on one affects another.

[a] Robert Curedale, *Design Thinking: Process and Methods*, 2nd ed., Topanga, Calif.: Design Community College Inc., 2016, p. 18.
[b] Improvising Design, homepage, undated.

solutions for pol-mil policy development, strategic and operational planning, programmatic development, and resource allocation. Approaches employed under design thinking are a SSM that incorporates systems thinking, critical thinking, and creative thinking. These methods and other visual methods promote competing views and facilitate insights. Reflecting a broader consensus within SOCOM about its importance, design thinking is a fundamental element of the Wargame Center approach toward wargaming.[130]

In addition to design thinking and critical thinking, the Wargame Center employs creative thinking and visual thinking in its process. Critical thinking is a process that imposes intellectual standards on the quality of thinking. It requires individuals to be open-minded and to gather, assess, and interpret information. Critical thinking also deals with points of view, quality of information, interpretation, inference, assumptions, implications, and consequences.[131] The center uses critical thinking processes to challenge the cognitive biases that stakeholders bring to decisions.[132] It also uses creative thinking, the process by which something new or useful is discovered and new solutions are created.[133] Additionally, visual thinking is an important component in a Wargame Center Senior Leader Seminar. As part of visual thinking, the center uses visual products to assist in conveying detailed patterns and complex systems to decisionmakers.[134]

Facilities

The Wargame Center conducts its wargames in a 3,500-square-ft facility. There are 74 seats arranged in elevated rows in a U-shape around a 20-ft-by-32-ft floor screen.[135] The floor screen is part of a virtual planning workspace system, with four downward-facing projectors that display two- or three-dimensional maps, imagery, pictures and videos. The virtual planning workspace is a flexible tool that can zoom in or out or pan around global map displays, showing overlays, range rings, heat maps, and other data points. This floor screen, large liquid crystal display screens at the front of the room, and multiple side monitors around the room create an environment for highly synchronized visualization displays, focused presentations, and facilitated discourse.[136] This space is pictured in Figure A.7.

---

[130] Telephone interview with SOCOM Wargame Center staff, November 2, 2016.

[131] UFMCS, 2016, p. 41.

[132] Email from SOCOM Wargame Center staff, October 5, 2017.

[133] Email from SOCOM Wargame Center staff, October 5, 2017. The Wargame Center staff referred to Harvard University's creative thinking course.

[134] Email from SOCOM Wargame Center staff, October 5, 2017.

[135] Telephone interview with SOCOM Wargame Center staff, November 2, 2016.

[136] Email from SOCOM Wargame Center staff, October 5, 2017.

Figure A.7
The U.S. Special Operations Command Wargame Center

SOURCE: USSOMC/ZUMA Press.

Process and Skill Sets

Reflecting senior leadership priorities, the Wargame Center focuses on policy and strategy decisions, emphasizing discourse, creative problem visualization, and innovative solutions. The center conducts a series of workshops ahead of an event to learn from SMEs, create empathy for the problem across diverse perspectives, and design innovative visualization products to better understand and address complex problems. This material is then presented to senior leaders utilizing short, powerful vignettes that illustrate select aspects of the problem set, with the aim of initiating informed facilitated discourse.

To design these events, the center uses a defined team consisting of a lead planner, a visual planner, and a game designer. The designer is an OR analyst with a specialty in design thinking. This core group facilitates the design of a Senior Leader Seminar with additional support from other elements of the Wargame Center, including IT support, audio support, and visual coding support. The center's design team then partners with an internal stakeholder, normally the SOCOM directorate with the most expertise and equity in the problem under analysis, and an external partner, which can be any organization or entity with expertise, equities, or interest. These external partners have included fellow CCMDs, other U.S. government agencies besides the DoD, and partner nations. The size of this combined design group varies by problem set.[137]

Wargames usually last one eight-hour day, broken into cycles of 15 minutes of problem comprehension and 45 minutes of discourse. The section on problem comprehension is designed, developed, and presented as a narrative/vignette akin to a TED Talk. A strong narration enhanced by visual products is meant to usurp the quintessential PowerPoint presentation style. Vignettes are informative, but short and concise, so that the time needed for their delivery is minimal, ceding most available time to the follow-on senior leader discourse instead. A facilitator presents critical questions related to the overall problem set, and/or each vignette, and loosely steers the discourse before moving to the next aspect/cycle of the problem. The game is geared to encourage fluid thinking and the exposure of differing perspectives on the same problem, including disagreements. Data capture mostly takes the form of note taking, and a considerable amount of time is spent after each game consolidating those notes, which then feed into an after action report that includes key insights, decisions, and a way ahead.[138]

Key Best Practices and Recommendations

The SOCOM Wargame Center specializes in wargames and seminars involving senior leaders with structured methods not used by other gaming centers. Although design thinking and structured techniques such as SSM are widely used in other contexts, SOCOM is the only one of the centers we contacted that brings these methods into wargaming.[139]

Other U.S. Gaming Organizations

Federally Funded Research and Development Centers—Wargaming Centers

The two defense-oriented federally funded research and development centers with wargaming activities are the CNA and RAND. We were not aware of major wargaming activities at either the Institute for Defense Analyses or the MITRE Corporation.

The Center for Naval Analyses
Overview

The CNA is a federally funded research and development center serving the Department of the Navy (including the U.S. Marine Corps) and other defense agencies. The center's efforts are defined by multidisciplinary, field-based, real-world, real-time research and analysis that combines observations of people, decisions, and processes.[140]

---

[137] Telephone interview with SOCOM Wargame Center staff, November 2, 2016.

[138] Telephone interview with SOCOM Wargame Center staff, November 2, 2016.

[139] Email from SOCOM Wargame Center staff, October 5, 2017.

[140] CNA, "CNA's Center for Naval Analyses," webpage, undated.

Wargaming at the Center for Naval Analyses

The CNA is well known in defense wargaming circles and conducts about 20 games a year, each typically lasting three to four days. Wargames cover a range of topics, including logistics, tactics, technology, ship-to-shore connectors, sexual assault, suicide prevention, climate change, and a variety of operational and organizational challenges. Wargame sponsors have included the Marine Corps, the Navy, the OSD, SOCOM, and a variety of functional and combatant commands.[141] CNA wargaming also appears to cut across all our categories of wargames.

Tools and Approaches

The CNA utilizes a number of tools and approaches for its wargames. These include technological enablers like Adobe Creative Suite/Creative Cloud, government-developed tools such as JWAM and SWIFT, matrix games, hex games, modified commercial games, and OR techniques. At the same time, as with many wargaming centers, the CNA stresses the importance of commercial board games as a source of inspiration for wargame mechanics. It notes that professional game mechanics are "an order of magnitude" simplified when compared with some commercial games.[142] Some of the CNA's tools and approaches are summarized in Table A.17.

Facilities

The CNA has three event-oriented spaces with different allowable classification levels. The largest area holds up to 150 people and can host wargames at the unclassified level. The space includes a large and completely reconfigurable area; a boardroom; three break-out rooms; a kitchen; a serving area; a conference center; and distinguished visitor or "admiral" spaces that allow for phone calls and conversations near the game. The CNA stressed several times during the course of our visit the importance of completely configurable spaces, saying that there was "no reason" not to have them. It also noted their modular furniture, as well as configurable and movable AV technology, IT, speakers, and landlines.[143] Although the facilities described above are multifunctional, the wargaming team does have a dedicated game room with a large-format printer, maps, and game construction materials.[144]

Process and Skill Sets

The CNA's wargaming teams consists of about 15 people who work on the various games. There are three wargame designers who are used across a variety of games. For a professional game, the CNA assigns a designer and a project director. The designer

---

[141] Interview with CNA staff, Arlington, Va., January 4, 2016.

[142] Interview with CNA staff, Arlington, Va., January 4, 2016.

[143] Interview with CNA staff, Arlington, Va., January 4, 2016. Wired landline phones are useful for communications between partitioned rooms.

[144] Interview with CNA staff, Arlington, Va., January 4, 2016.

Table A.17
Center for Naval Analyses Wargaming Tools and Approaches

Tool or Approach	Description	Usage
Adobe Creative Suite/Creative Cloud	Software suite of graphic design, video editing, and web development products that has become the industry standard across various fields.	Used to edit and create news broadcasts, posters, and other injects that propel the game's narrative.
Board and card games	Games intended to introduce abstract concepts like deterrence and cooperation.	Used for various wargame design purposes.
Configurable physical gaming space	Large, configurable space for gaming. Ideally, the space can be subdivided into space of different sizes, have furniture that can be easily moved, and allow for custom AV and IT configurations.	Used for diverse games including computer-assisted games, manual board games, and matrix- and seminar-style games.
Harpoon	A game for two to eight players that covers various aspects of maritime combat including surface, subsurface, and air engagements.	Used in conjunction with other documentation for up to date weapon and platform ratings allow for relatively realistic evaluations of scenarios.
JWAM	A manual, computer-aided, time-step, human-in-the-loop, force-on-force simulation methodology developed over ten years at the CAA.	Used for evaluating and comparing COAs to support CCMD operational plans, and for defense planning scenario development.
Large-format printers	Printers that allow a maximum width of 18–60 inches.	Large-format printers are useful for printing maps, posters, and various visualization aids for use during games.
Matrix games	Means of structuring gameplay between multiple teams that enables players to weigh in on the likely outcome of other teams' actions.	Mainly used for adjudication, visualization, and data collection. (Not used very often at TRAC.)
Modified commercial board games	Commercially available games for educational or analytical purposes. For all but the simplest games, modification is generally needed to make rule sets more accessible and create a playable game to address the educational requirement.	Used for education and experiential learning on topics including COA development, principles of COIN and ethics and decisionmaking.
Methods of Operations Research and Search and Screening[a]	General-purpose textbooks that provide fundamental OR methods with various military applications.	Used for wargame research and design and provides methods for measuring the effectiveness of certain tactics.
Spreadsheet tools	Excel-based tools that run in the background during games.	Used mostly for bookkeeping and accounting.
SWIFT	A platform for capturing gameplay, adjudication, data analysis, and constructing COPs.	Used for bilateral and multilateral wargames, logistics, personnel recovery, cyberoperations, and exploring future conflicts with Russia.

Table A.17—Continued

Tool or Approach	Description	Usage
Tablets	Touch screen enabled tablet computer useful for media display.	Used for scenario presentation.
Videos/video studio	Ranging from a studio for live video production to camcorders and computers with editing software, several centers stress the utility of being able to produce "new" video to deliver game scenario briefings and injects for heightened impact.	Used for introduction briefings of game scenarios and scenario updates.

[a] Philip M. Morse and George E. Kimball, *Methods of Operations Research*, Mineola, N.Y.: Dover Publications, 2003; Bernard Osgood Koopman, *Search and Screening: General Principles with Historical Applications*, Alexandria, Va.: Military Operations Research Society, 1999.

handles the scenario, game concept, and execution. The project director manages the process, data collection and analysis plan, event logistics, project management, and postgame analysis for report writing. There is also a project manager when there are significant administrative tasks, such as travel bookings. The entire process for a game may take three to six months or more depending on the scope and intricacies of the game.[145]

Historically, the CNA has hired hobby gamers in game designer roles, and taken wargamers and used them as graphic artists. Preferring more flexibility in game adjudication and design, the CNA does not typically design game mechanisms requiring software programming.[146]

Key Best Practices and Recommendations

Several times the CNA emphasized the importance of configurable spaces and the importance of having a wide variety of commercial game mechanics from which to draw inspiration for professional games. It also emphasized the importance of the narrative aspects of games, both in advancing and re-creating the narrative in a game.

RAND Corporation

Overview

RAND is a nonprofit research organization that develops solutions to public policy challenges based on objective and independent analysis. RAND has used gaming to support its analytic efforts since 1947, not only in support of DoD but also in all other policy issue areas like health and infrastructure. Within the defense sector, RAND supports a wide range of sponsors from the armed services to NATO on topics like

---

[145] Interview with CNA staff, Arlington, Va., January 4, 2016.

[146] Interview with CNA staff, Arlington, Va., January 4, 2016.

North Korean scenarios, Russian aggression in the Baltics, force structure and capabilities, and emergency response.[147] While not responsible for day-to-day game design or client management, the RAND Center for Gaming under the Pardee RAND Graduate School is designed as a hub for the cultivation and circulation of best practices throughout the RAND community.

RAND wargames vary in size, topic, and complexity. RAND's game design teams generally consist of both specialists in game design and execution, as well as a range of SMEs.

Wargaming at RAND

RAND games are generally analytical and focus on the operational or strategic level, though a small number of more tactical games do occur. They are used to examine warfighting concepts, train and educate commanders and analysts, explore scenarios, and assess how force planning and posture choices affect campaign outcomes. RAND has developed and can execute various types of wargames, including scenario exercises, tabletop MAPEXes, and computer-supported exercises.[148] The duration of a RAND wargame can vary from a few hours to many days depending on the scope of the problem and the sponsor.

Tools and Approaches

Due to the variety of wargames RAND conducts, it utilizes a plethora of tools and approaches. Depending on the sponsor's objectives, RAND tailors its approach and toolset to each wargame. For instance, the Day After methodology, a strategic planning exercise, was originally designed twenty-five years ago as a way to explore a non-apocalyptic use of nukes. Over time the methodology has been used to explore critical infrastructure, money laundering, informational warfare, and cyberoperations, among other subjects. Other RAND methods, such as the RAND Framework for Live Exercises (RFLEX), are designed to support games focused on major combat operations. Other games have been designed to look at problems such as security force assistance, U.S. operations to counter the Islamic State of Iraq and the Levant (ISIL), or whole-of-society responses to cyberattacks.

RAND's wargaming approaches and tools are summarized in Table A.18.

Facilities

RAND does not have dedicated spaces for gaming, but does use multipurpose classified and unclassified venues for wargames, enabled with VTC and various IT services. RAND also conducts wargames for its sponsors at their respective locations, domestically and internationally. RAND has wargames based in its various offices in Cambridge, England; Pittsburgh; Santa Monica, California, and Washington, D.C.

---

[147] Pardee RAND Graduate School, "Center for Gaming," webpage, undated.

[148] Pardee RAND Graduate School, undated.

Table A.18
RAND Wargaming Tools and Approaches

Tool or Approach	Description	Usage
Adobe Creative Suite/ Creative Cloud	Software suite of graphic design, video editing, and web development products that has become the industry standard across various fields.	Used to design maps and other game materials.
Board and card games	Games intended to illustrate core ideas or processes.	ISIL game used by RAND for educational purposes.
Combat Operations in Denied Environments	A suite of four models that identify detailed lists of combat support requirements for various basing postures and scenarios, optimal locations for war reserve material, and investments in defense resources and their effect on the Air Force's ability to conduct operations in denied environments.	Used for decision support for the Air Force in determining the required resources for various basing strategies.
Day after methodology	A series of multistage case study exercises designed to focus experts and decisionmakers on the concrete issues of a particular policy dilemma.	Currently used for critical infrastructure and disaster scenarios for the Department of Homeland Security, public health emergencies for local and state governments, and exploring the utility of new technology.
Large-format printers	Printers that allow a maximum width of 18–60 inches.	Used to create graphics as an analytic tool to inform decisions about concept development, acquisition, force design, force mix, and tactics, techniques, and procedures development.
Matrix games	Means of structuring gameplay between multiple teams that enables players to weigh in on the likely outcome of other teams' actions.	More structured variant of a seminar-style game, particularly useful when gaming emerging topics with limited expertise to inform adjudication.
RFLEX	Hex-game-based "live wargame" methodology consisting of blue and red teams usually (but not always) working around a common map with a typically transparent adjudication process.	Used to reveal decision points, broad strategies, and how participants think about and deal with unexpected outcomes. Generally used to examine operational-level problems, such as conflicts with near peers and regional powers.
SATs	Techniques that the intelligence community uses to improve intelligence analysis.	Used to support Marine Corps intelligence and wargaming.

Process and Skill Sets

Due to the wide variety of games, sponsors, and groups within RAND that wargame, it is difficult to pinpoint any process as a particular RAND process to wargaming. Also, due to its matrixed organization and internal labor market, where project staffing is decentralized, there is no particular group of people who act as RAND's dedicated wargamers.

While that is highly unusual compared with other organizations that wargame, RAND's wargaming, including its approaches, tools, and staffing, are determined at the individual project level. That said, RAND wargame projects do have game designers, SMEs on both the regional and functional focus of the games, note takers, facilitators, and analysts who support game execution and analysis.

Key Best Practices and Recommendations

Over the past three years RAND has expanded the types of games that it uses. In addition to the standard hex game first developed by RAND in the 1950s and the Day After methodology that was a hallmark of RAND wargaming in the 1990s, RAND has added matrix gaming and games using SATs, and is looking into VR augmentation of board games.

RAND has benefited from being able to draw on staffs with diverse areas of expertise and a sustained research program to inform its games. For example, many of the more structured adjudication models currently used at RAND were informed by more narrow M&S efforts in the particular topic at hand.

Allied Gaming Centers

We were also able to speak with wargamers from allied defense ministries in Australia, Canada, and the United Kingdom. The U.S. military is significantly larger than that of its allies, and the scale and scope of U.S. defense wargaming is also comparatively larger. We have not tried to apply the wargaming categories used elsewhere in this report to these allies efforts because we do not want to force-fit allied gaming concepts into categories that have been developed from a U.S. defense organizational construct.

The Defence Science and Technology Laboratory, UK Ministry of Defence
Overview

The Dstl contains the largest single concentration of MOD professional wargamers. As such, the Dstl provides independent, impartial S&T expertise to MOD and various other British government agencies. In terms of wargaming, the Dstl provides a wide range of capabilities, tools, and types of wargames. Its wargaming capability is centralized within the Wargaming Team, which is part of the Defence and Security Analysis Division, based at Dstl, Portsdown West. It currently has approximately 30 full-time staff and access to additional expertise across the Dstl, the wider MOD, academia, and industry.[149]

---

[149] Email from Dstl staff, February 16, 2018.

Wargaming at the Defence Science and Technology Laboratory

The Dstl views wargaming as a process of competitive challenge and creativity delivered in a structured and umpired/adjudicated format. It has adversarial elements (these can be challenging and "oppositional" factors, as well as hostile actors); it considers dynamic events (thus illuminating any issues created by constant change); and it is driven by player decisionmaking. The purpose of all wargaming is to immerse participants in an environment with the required level of realism to improve their decisionmaking; it is about the participants, the decisions they make, the narrative they create, shared experience, and the lessons that they take away. A wargame is a simulation—it is a representation of real life. However, it should not be confused with constructive simulation models or synthetic environments that may or may not be used to support a wargame.[150]

Dstl wargames are internally classified into several categories: tactical, operational, strategic, seminar, thematic, experiential, and wargames exploring the future (roughly 20 years ahead). Such wargames are used to explore tactical, operational, strategic, and grand-strategic issues across the full spectrum of public, private business, security, emergency services, and military activity. They can be used to identify emerging issues, test hypotheses, examine alternate options and ways of thinking, and highlight the potential consequences of choices—all set against the actions of a dynamic adversary or situation.

Wargames can provide

- an opportunity to take risk without risking lives or disrupting business continuity
- a cost-effective way to practice command and exercise staff and management skills
- exposure to friction and uncertainty, including adaptive, thinking adversaries, allies, competitors, and stakeholders
- a mechanism for exploring innovation in both warfare and business.

The process of wargaming is at least as important as the results. This is obviously true for wargaming in support of training and education, but is equally true when used to support planning and decisionmaking. It is crucial that defense personnel are involved in the whole process of wargame design, execution, and exploitation. Wargaming needs to be part of our DNA and processes.

Furthermore, wargaming enables the integration of different methods, tools, and techniques (both quantitative and qualitative) with a human element and thereby creates a capability greater than the sum of its parts. However, wargaming should not be considered a panacea. A single wargame event will rarely provide a definitive, robust, quantitative answer to a particular problem. Wargaming provides greatest util-

---

150 Telephone interview with Dstl staff, September 8, 2017.

ity when used iteratively within a wider decisionmaking process. It is complementary to and complemented by a range of other approaches, including red teaming and net assessment.[151]

Tools and Approaches

Due to the diverse types of Dstl wargames, its wargaming program utilizes a plethora of tools and approaches. A sampling is provided in Table A.19.

Key Best Practices and Recommendations

The Dstl's wargaming program is bolstered by its diversity in approaches and tools. Its wide range of clients and sponsors requires a high degree of flexibility and adaptability to its wargames. This is in contrast to service-oriented wargaming programs or

Table A.19
Defence Science and Technology Laboratory Wargaming Tools and Approaches

Tool or Approach	Description	Usage
Close Action Environment (CAEn)	A multisided, computerized wargame and simulation capable of representing all-arms close-combat battle from individual soldier or platform to company group level. CAEn can model both highly detailed rural and urban terrains in a variety of weather and lighting conditions. CAEn is most commonly run as a human-in-the-loop, closed wargame, with commanders only knowing what their units have detected.	CAEn is primarily used to investigate the impact of changes to scheme of maneuvres, capability and force structure on the ability of Coy and below forces to complete the mission.
Configurable physical gaming space	Large configurable space for gaming.	Used as a venue for diverse games including computer assisted, manual board games, and matrix- and seminar-style games.
Dilemma analysis (also called conformation analysis)	Software method to structure multiparty conflicts (including negotiations and more active conflicts) by collecting information on priorities, statements from the actors about their preferences, the potential results of not resolving the dispute, and where distrust exists between players.	Currently being experimented with in shaping game designs.
Matrix games	Means of structuring gameplay between multiple teams that enables players to weigh in on the likely outcome of other teams' actions.	Used as a substitute for seminar-style games, particularly when gaming emerging topics with limited expertise to inform adjudication.

---

[151] Email from Dstl staff, February 16, 2018.

Table A.19—Continued

Tool or Approach	Description	Usage
MaGCK Matrix Game Construction Kit	Commercially available matrix game design kit codeveloped by the Dstl.	Used to create matrix games.
Peace Support Operations Model (PSOM)	A computer-assisted wargame designed to represent the full range of civilian and military activity in a stabilization scenario. The wargame provides a method to analyze the stabilization problem space within the context of a wider cross-government response.	Decision-support tool used for examining questions relating to stabilization, COIN, and irregular warfare operations.
RCAT	A manual simulation used to support a wargame. RCAT can support a range of requirements, from facilitating insights at theme-focused workshops to wargaming analysis of campaigns using tailored levels of complexity. The toolset consists of a set of baseline mechanisms and rules, and some extended and enhanced mechanisms in certain topics (e.g., cyberoperations).	Various uses, including for force structure, force development, and training.
Seminar games	Games that enable an open-ended, argument-based discussion between experts to elicit opinions and judgments. Players are immersed in a context, asked to make decisions, and then face the consequences of those decisions. Adjudication can be semirigid but tends toward free. Seminar games are usually conducted in small groups; hence the name.	
Technical decision support wargame	Card-based game used to identify future capabilities by structuring selection of packages of technologies to form capabilities and explore the capabilities use under different vignettes.	Mainly used for game design, COPs, and adjudication.
Wargame Infrastructure and Simulation Environment: Formation Wargame	Computer-based, human-in-the-loop wargame, and constructive simulation for battle group to division tactical actions, including air and maritime support. It is a fully closed wargame, with commanders only knowing what their forces have reported to them.	Testing the impact of changes to scheme of maneuvers, capability, and force structures on the ability of forces to complete the mission.
Wargaming Handbook	Handbook published in August 2017. The scope is to introduce wargaming fundamentals; types, variants, and contexts; processes; case studies, further reading, and defense points of contact.	To "provide context and guidance for wargaming. It is designed principally to introduce the topic; it is not a detailed manual or practitioner's technical guide."[a]

NOTES: [a] UK MOD, Development, Concepts and Doctrine Centre, *Defence Wargaming Handbook*, Swindon, England: UK Ministry of Defence, August 2017, p. v.

wargaming programs aimed exclusively at senior leadership.[152] The Dstl supported pub-lication of the Development, Concepts and Doctrine Centres *Wargaming Handbook* in August 2017. This is the first publication of its type in UK defense, and its purpose is "to provide context and guidance for wargaming."[153] Figure A.8 shows the *Wargaming Handbook*. The Dstl also supports Connections UK, a professional wargaming confer-

Figure A.8
The UK *Wargaming Handbook*

SOURCE: UK MOD, 2017.

---

[152] Telephone interview with Dstl staff, September 8, 2017.

[153] UK MOD, 2017, p. v.

ence attended by a broad range of delegates from the military, defense analysis, business, emergency services, humanitarian organizations, and academia.[154]

Center for Operational Research and Analysis, Defence Research and Development Canada
Overview

As one of seven research centers within Defence Research and Development Canada, CORA's core mission is "to inform and facilitate the decision-making process of the Canadian Government defence and security community in an objective, timely, credible and scientifically rigorous way. As such, CORA provides senior decision makers with quantitative and qualitative decision support and expert advice in the broad areas of operational research, strategic analysis, social sciences research, and scientific and technical intelligence."[155]

Wargaming at Center for Operational Research and Analysis

CORA's wargaming capacity and capabilities have become smaller in recent years. During the 1990s, CORA fielded a robust computer-based wargaming program, which included the use of software tools such as CAEn and JANUS and a rigorous wargaming schedule supporting force and tactics development. With the end of the Cold War, and shifting institutional priorities, wargaming capabilities at CORA steadily diminished over the years. Currently, CORA's wargaming program is informal and consists of seminar and tabletop games used for policy development and analysis, support to strategic planning, and tactics development.[156] Approximately ten staff members across CORA are involved who conduct game planning, execution (including facilitation), and analysis. Most games are on the scale of a half dozen players but larger wargames, consisting of up to ten to 20 players, are run every few years. Most CORA wargames last only a day, but occasionally there is two- to three-day game.[157]

Tools and Approaches

Limitations on CORA's staff and resources constrain the tools and approaches utilized in its wargaming program. At present the center does not leverage software tools in its wargames. At CORA, the dominant approach toward wargaming consists of facilitator-led matrix- and seminar-style games.[158] Typically, each wargame will have staff notes, player surveys, and some form of player hot wash briefing, all of which gets

---

[154] Email from Dstl staff, February 16, 2018.

[155] Email from CORA staff, September 20, 2017.

[156] CORA defines seminar games as a facilitated and structured discussions, in contrast with tabletop games that use game materials such as maps or other playing aids that serve to define the game scenario and constraints for the players.

[157] Telephone interview with CORA staff, August 23, 2016.

[158] Matrix-style games are considered a form of tabletop game by CORA.

Table A.20
Center for Operational Research and Analysis Wargaming Tools and Approaches

Tool or Approach	Description	Usage
Matrix games	Means of structuring gameplay between multiple teams that enables players to weigh in on the likely outcome of other teams' actions.	Used as an alternate for seminar-style games, particularly when gaming emerging topics, planning scenarios, and policy issues.
Seminar	Structured discussion on a specific scenario or problem set, leveraging subject matter expertise.	Used for exploring strategic, operational, or policy impacts of a specific vignette and for exploring new ideas, concepts, and/or intangibles at the operational/strategic level.

compiled into a final report. However, the center does not have a structured framework or collections methods in its wargaming.[159] The tools and approaches utilized by CORA are summarized in Table A.20.

Process and Skill Sets

The majority of CORA's staff come from OR backgrounds with quantitative skill sets. However, for its policy wargames, CORA incorporates qualitative expertise, usually in the form of historians or political scientists. The staff generally do not have formal training in wargaming. Nevertheless, staff members have supplemented their understanding of wargaming by attending MORS and other wargaming conferences. In recent years CORA adopted matrix wargaming and is currently experimenting with the RCAT methodology after its British counterpart introduced CORA to this approach.[160]

Key Best Practices and Recommendations

CORA recommends matrix gaming as a very flexible methodology that can be adapted to a wide variety of applications and does not require large resource investments. Matrix gaming has grown increasingly popular over the years, as demonstrated by the adoption of the method by academia and many countries, including Canada.

Defence Science and Technology Joint and Operations Analysis Division, Australian Department of Defence
Overview

The Joint and Operations Analysis Division resides within its larger parent organization, DST. Broadly speaking, the DST's mission is to provide technological and strategic knowledge in transforming the Australian Defence Force and Australia's national security. Specially, the Joint and Operations Analysis Division specializes in the development of future capability through the evolution of joint concepts, future force structure, capability development for individual services, and opportunities to leverage dis-

---

[159] Telephone interview with CORA staff, August 23, 2016.

[160] Telephone interview with CORA staff, August 23, 2016.

ruptive technologies. The division's topical areas include system of systems engineering, concepts for networked operations, modeling, simulation, and experimentation.[161]

Wargaming at the Joint and Operations Analysis Division

In terms of wargaming, Australian Army headquarters serves as the principal customer.[162] Therefore, DST's wargames tend to concentrate on the modernization of land forces focusing in areas including force structure, environmental challenges, adversarial concerns, future capabilities, and gaps in the current force. DST wargames range from the tactical level of the individual soldier to the operational level, where the battle group is the largest aggregation. Each year's slate of wargames and their topics are based on a series of questions proposed by the Australian Department of Defence. Thus, DST tends to leverage wargaming as a method testing analytical campaigns to test military plans. The number of wargames DST executes on an annual basis fluctuates with institutional priorities and needs. Typically, DST wargames focus on future scenarios, roughly ten to 15 years into the future, and usually involve force structure, environmental, adversarial, and capabilities concerns.[163]

Tools and Approaches

DST uses a handful of different tools and approaches, but mainly utilizes jSWAT2, a digitalized representation of a manual, paper-based seminar wargame, for its games. The program allows players to contextualize units within the wargame in time and space. Moreover, players can configure the units in flexible ways while simultaneously capturing the data from gameplay. A common wargaming approach used with DST is break point analysis. This process is focused around finding break points through the variation of key contextual variables.[164] Organizationally, DST emphasizes flexible decisionmaking and assessing the consequences of specific decisions within its wargames.[165] A summary of tools and approaches utilized by DST is provided in Table A.21.

Facilities

DST's wargaming program has a dedicated facility with a number of linked rooms. Some of the rooms are configurable in terms of shifting tables, chairs, and screens. However, in practice, DST typically travels to the client and uses training spaces

---

[161] Australian Department of Defence, Science and Technology Group, "About DST," webpage, undated.

[162] Here the term *wargaming* refers to Analytical Seminar Wargaming. Other examples of wargaming within the Joint and Operations Analysis Division and DST include virtual and constructive simulation and red teaming.

[163] Telephone interview with DST staff, November 23, 2016.

[164] These variables are drawn from the broad dimensions of nature, diversity, and intensity of threat; operational partnerships; sociopolitical issues; complex human terrain; limits, capabilities, and options of forces; and complex physical environment. See Brandon Pincombe et al., "Ascertaining a Hierarchy of Dimensions from Time-Poor Experts: Linking Tactical Vignettes to Strategic Scenarios," *Technological Forecasting and Social Change*, Vol. 80, No. 4, 2013, pp. 584–598.

[165] Telephone interview with DST staff, November 23, 2016.

Table A.21
Defence Science and Technology Wargaming Tools and Approaches

Tool or Approach	Description	Usage
Dynamic morphological exploration (DME) tree	Add-on approach developed by DST to general morphological analysis (GMA) that creates a tree mapping of optimal search paths through a morphological space.	Used to examine combat vehicle variations.
jSWAT2	Computer-based tool for a seminar wargaming environment, which includes a planning environment to aid in creating synchronization matrices and a simulation used to adjudicate maneuver, logistics, combat, and intelligence gathering.	Used to explore the Australian Army Experimental Framework, Air Power.
Zing Portable Team Meeting System	Software and hardware package designed for real-time collaboration, where users can anonymously project comments on a screen so they are not spoken over.	Used during wargames to allow members to equally contribute.

provided by the Army. For enhancements in facilities for wargaming, DST suggested a series of rooms connected electronically for data sharing, projection and recording capabilities, and potentially the use of one-way mirror rooms, which will allow for unobtrusive observation of gameplay and discussion.[166]

Processes and Skill Sets

DST's principal wargaming staff consists of 14 members with various backgrounds, such as in mathematics, computer science, and history. However, most do not have formal OR degrees. Nevertheless, DST argues that its smaller size and variety of backgrounds enables organizational flexibility and agility.[167]

Key Best Practices and Recommendations

The most notable practice of DST wargaming is its use of jSWAT. Despite its simplicity, the program enables substantial flexibility in use for wargamers and players. If properly utilized, jSWAT can serve as powerful tool for analysts, players, and facilitators. The program maps branches in each decision while allowing players to explore other possibilities at different decision points. Additionally, jSWAT is useful for postgame analysis and after action reports, as it visualizes specific talking points and context of decisions, as well as capturing some of the thought process of the players throughout gameplay. DST characterized the program by saying, "The cleverness is in its simplicity."[168]

---

[166]   Telephone interview with DST staff, November 23, 2016.

[167]   Telephone interview with DST staff, November 23, 2016.

[168]   Telephone interview with DST staff, November 23, 2016.

Catalog of Wargaming Tools and Approaches

OAD tasked the RAND Corporation with compiling a catalog of wargaming tools and approaches used by a number of different wargaming centers in 2016, and we provided an initial list in 2017. The catalog of 77 tools in this appendix has been updated since that initial list and reflects information through 2017 and, in some cases, early 2018. The catalog includes entries for the major wargaming tools and approaches that wargaming centers brought up during the course of our 2016–2017 visits and discussions with them. This is not an exhaustive list, and the wargaming centers contacted for the project are by no means all of the organizations that wargame.

These tools and approaches span the range from low-tech to computationally intensive. They support a range of wargame-related activities, including knowledge management, planning, visualization, adjudication, data collection, and data analysis. It is difficult to fully convey the variety of tools employed in wargaming, or the wide range of wargame types used by the community. Listing individual catalog entries tends to take these tools and approaches out of the organizational contexts and processes in which they are employed, which is why we also list each wargaming center's major tools and approaches in Appendix A.

The purpose of this appendix is to provide wargamers and those seeking to develop wargaming capabilities with an idea of what defense wargamers currently use in practice. Appendix B differs from the existing OSD wargaming repository's tool database in several ways. One is that the information is entirely unclassified and it does not cover any classified information on tools. (The OSD repository exists on SIPRNet.) Another difference is that this appendix uses a somewhat broader definition of "tool" than the online repository and includes entries that are unlikely to make it into the repository. Another distinction is that this appendix was generated from the organizations that we contacted and does not cover the entire list of organizations that may have added information to the online tools database. We have no doubt that the list we provide would have been longer given the time to conduct interviews at more organizations. However, it covers allied wargaming centers that do not currently enter information into the OSD repository. Finally, this appendix contains some additional details specifically requested by the Marine Corps, such as tool requirements, that are not found in the repository.

Adobe Creative Suite/Creative Cloud
- Overview
  - Center: CNA
  - Tool/approach: Adobe Creative Suite/Creative Cloud
  - Description: A software suite including Adobe Illustrator and Photoshop, of graphic design, video editing, and web development products that has become the industry standard across various fields. Illustrator is a vector-based image tool that can be used to create resolution-independent graphics that are easily modifiable and scalable. Photoshop, a commonly used application in the Creative Suite, can be used to manipulate and edit images, enhance photos, perform graphic and web design, and edit video.
  - Government sponsor: None
  - Developer: Adobe
- Information
  - Focuses: Visualization and presentation
  - Operational levels: All
  - Purpose: Facilitates the creation of high-quality graphics and videos that can be used during the course of a game to increase realism and improve the player experience
  - Forces: All
  - Current uses: Some gaming centers use these products to edit and create news broadcasts, posters, and other injects that propel the game's narrative
  - Limitations: The stand-alone version of Adobe Creative Suite has recently been retired and is no longer available for purchase. Instead, Adobe provides Creative Cloud, which is an internet-based subscription service that provides access to the same tools. Adobe is willing to work with large organizations that have restrictions on the use of web-enabled products.
- Requirements
  - Facilities: N/A
  - Equipment: Adobe products can be run on most current desktop computers
  - Personnel: One person to design and edit graphics/video products
  - Required training: Adobe products are professional-grade tools with a wealth of options and potential complexity. However, because they are highly regarded commercial products, numerous training resources both online and in person are available from a variety of sources.
  - Cost: $69.99 per month for cloud access to the full creative suite; $29.99 per month for access to individual applications

Notes:
- *See also* the Videos/Video Studio entry.

Advanced Warfighting Simulation
- Overview
  - Center: TRAC
  - Tool/approach: AWARS
  - Description: A typically closed-loop (no-human-interaction), deterministic (expected-value), unit-level simulation representing ground and amphibious warfare with variable resolution down to the platoon level
  - Government sponsor: U.S. Army
  - Developer: TRAC
- Information
  - Focus: M&S of ground and amphibious warfare
  - Operational levels: From the platoon to the joint task force
  - Purpose: An analytic tool to inform decisions about concept development, acquisition, force design, force mix, and tactics, techniques, and procedures development. AWARS allows analysts to understand how a capability or concept contributes to military operations.
  - Forces: Ground and amphibious; joint assets are represented
  - Current uses: Army acquisition decision support and doctrine development. AWARS is used for the comparative analysis of alternative COAs, specific systems/system attributes, and/or concepts—and how the alternatives contribute to military operations relative to a baseline. AWARS is not used for predictive analysis.
  - Limitations: TRAC uses certified performance data from Army Materiel Systems Analysis Activity. Operational data comes from SMEs in the CCMDs, the Joint Services, and TRADOC.
- Requirements
  - Facilities: "Battle lab" for AWARS
  - Equipment: Multiple computers
  - Personnel: Twenty dedicated AWARS coders at TRAC
  - Required training: Significant; dedicated game team to use AWARS

Notes:
- Often used for brigade, division, and corps-level analyses
- Traditionally used by TRAC-FLVN for operational-level wargames and analyses
- TRAC has 20 coders dedicated to AWARS and a "battle lab."

Air Force Materiel Command Wargaming Course
- Overview
  - Center: AFMC
  - Tool/approach: AFMC wargaming course

- – Description: A for-official-use-only wargaming course developed by AFMC to instruct its personnel on how to execute a wargame
  - – Government sponsor: U.S. Air Force
  - – Developer: AFMC
- • Information
  - – Focus: Wargame education
  - – Operational levels: Tactical and operational
  - – Purpose: To serve as an educational foundation for wargaming
  - – Forces: Air
  - – Current uses: Utilized by Air Force personnel to learn the basics of designing, executing, and analyzing a wargame
  - – Limitations: Does not have a well-defined doctrine or teaching method
- • Requirements
  - – Facilities: No specific facilities required. Considering using the National Air and Space Intelligence Center Warfighting Center.
  - – Equipment: Unclear
  - – Personnel: Unclear
  - – Required training: Unclear

Notes:
- • Since AFMC is largely focused on its service Title 10 wargames and its related FAST wargames, there is limited value of its wargaming course to other services. Unlike the NPS, which provides a more generalist approach to wargaming education, AFMC's wargaming course does not seem to possess the same approach.

Analysis of Competing Hypotheses
- • Overview
  - – Center: NWC
  - – Tool/approach: ACH
  - – Description: ACH is an analytic process to systematically enumerate and evaluate a complete set of hypotheses based on all available evidence. Developed for the intelligence community in the 1970s, ACH was designed to aid intelligence analysis and has since been implemented in software form for wider use. ACH forces individuals to identify all potential hypotheses, list evidence and arguments, and organize the information into a matrix form so that the individuals can examine all hypotheses against each piece of evidence. This method provides auditability and a means for overcoming certain cognitive biases.
  - – Government sponsor: N/A
  - – Developers: Multiple implementations of ACH in software, including open-source versions (e.g., Competing Hypothesis), versions by SSS Research Inc. (DECIDE), and the Palo Alto Research Center (ACH 2.0)

- Information
  - Focus: Analysis
  - Operational levels: All
  - Purpose: According to one reviewer, the software "takes you through a process for making a well-reasoned, analytical judgment. It is particularly useful for issues that require careful weighing of alternative explanations of what has happened, is happening, or is likely to happen in the future. ACH is grounded in basic insights from cognitive psychology, decision analysis, and the scientific method."[1] Although ACH can be done with paper and pencil, software implementations allow for better data management and visualizations. In wargaming, ACH can be beneficial for controversial issues where analysts can create an auditable trail that details their considerations and the reasons for their conclusions.
  - Forces: All
  - Current uses: Navy wargaming analysis
  - Limitations: Open-source software is not officially supported and other software solutions appear relatively old (ca. 2006)
- Requirements
  - Facilities: Minimal facility requirements; workspace for computers only
  - Equipment: Competing Hypothesis, the open-source software, runs as a web service and therefore requires some type of server, either Apache or Windows, connected to a computer with a web browser installed. However, because ACH is also a methodology, it does not necessarily require anything more than paper and pencil.
  - Personnel: Individuals with experience in the ACH methodology
  - Required training: Training support offered by third-party companies, including Pherson Associates, which offers consulting, workshops, and courses over a range of analytic activities
  - Cost: Free

Notes:
- Identified by both the intelligence community as a structured analytic technique and UFMCS as a red teaming technique.

The Applied Critical Thinking Handbook
- Overview
  - Centers: UFMCS, CSL
  - Tool/approach: *The Applied Critical Thinking Handbook*
  - Description: Handbook of structured group methods aimed at reducing bias, mitigating groupthink, and encouraging critical thinking

---

[1] Alexandra Vaidos, review of ACH, Softpedia, undated.

- – Government sponsor: UFMCS (under TRADOC)
- – Developer: UFMCS
- Information
  - – Focuses: All
  - – Operational levels: All
  - – Purpose: Framing questions, generating approaches, evaluating potential approaches
  - – Forces: All
  - – Current uses: Utilized by Army, Marine Corps, SOCOM, U.S. Customs, and Border Patrol
  - – Limitations: Methods are chosen and taught with U.S. military culture in mind. Red teaming techniques are not currently used often in wargaming.
- Requirements
  - – Facilities: Minimal
  - – Equipment: Minimal
  - – Personnel: Red team leader or facilitator
  - – Required training: At UFMCS, 9 or 18 weeks to qualify as a red team leader, six weeks to be a red team member; other customizable training is available off-site to cover different red teaming components

Notes:
- UFMCS, *The Applied Critical Thinking Handbook*[2]
- Additional red teaming handbooks, guides, and articles are available at the UFMCS website
- Red teaming is also practiced by allies, including the United Kingdom, which has an established program based at the Development, Concept, and Doctrine Centre
- The Marine Corps was an early adopter of red teaming. However, several key red teams, such as the commandant's red team, were recently disbanded. As a result, seasoned red teamers with high-level experience may be available, but only for a limited time.
- LtGen Paul Van Riper (Ret.) and Ben Jensen of MCU may be resources, particularly on the intersection of complexity studies and red teaming
- To be successful, red teams require protection from superiors to preserve their special function. Red team members should not be short-term participants, nor should red teaming be their permanent job.
- Training costs would only be travel expenses for anyone coming from the Marine Corps. Eighteen-week leader course is for SOCOM only.
- The director of UFMCS spoke at an October 2016 MORS special meeting panel and command red teams

---

[2] UFMCS, 2016.

- Intelligence community structured analytic techniques, red teaming, "soft operational analysis" or judgment-based operational analysis, and problem-structuring methods from OR share common methods.

ArcGIS

- Overview[3]
  - Center: SOCOM
  - Tool/approach: ArcGIS
  - Description: Application that offers a unique set of capabilities for applying location-based analysis to business practices. ArcGIS enables the analysis and visualization of data through contextual tools, offering deeper insights. The application also enables sharing and collaboration functions.
  - Government sponsor: N/A
  - Developer: Esri
- Information
  - Focus: Visualization
  - Operational level: Any
  - Purpose: Visualization
  - Forces: Any
  - Current uses: ArcGIS creates deeper understanding, allowing analysts to quickly see where things are happening and how information is connected
  - Limitations: Concerns about limited built-in symbols
- Requirements
  - Facilities: N/A
  - Equipment: ArcGIS desktop software
  - Personnel: Minimal
  - Required training: Minimal

Athena

- Overview
  - Center: TRADOC G2
  - Tool/approach: Athena
  - Description: Computerized simulation of sociocultural environments
  - Developer: National Aeronautics and Space Administration Jet Propulsion Lab
- Information
  - Focuses: Counterterrorism, economic, interagency, policy, and stabilization
  - Operational levels: Operational and strategic. In theory, Athena could go lower because of a scalable system, but the number of interacting elements becomes unwieldy, and reliable data become very difficult to find to populate the model.
  - Purpose: Adjudication, analysis, and training

---

3 A free trial and demo can be found at ArcGIS, homepage, undated.

- – Forces: Any, but the focus on population makes it particularly pertinent for ground forces
- – Current uses: SOCOM and Central Command analysis and game support
- – Limitations: Fidelity of model limited by availability of input data
- Requirements
  - – Facilities: None
  - – Equipment: Standard computer (does not require server)
  - – Personnel: G2 may be able to support; needs SME input to the model, as well as the team to run it. Requires familiarity with sources of sociocultural data helpful.
  - – Required training: Learning to interpret how the model runs generally takes a few weeks. System requires time to learn and master, which will result in slower modeling. Building new capabilities can be done in Python.

Notes:
- Model is government rights software with an Army Network Command Certificate of Networthiness, including approval for SIPRNet—code is available on GitHub, complier from JHU-APL
- Political, Military, Economic, Social, Information, Infrastructure, Physical Environment, and Time (PMESII-PT) variables as policy levers to affect Diplomatic, Information, Military, Economic, Financial, Intelligence and Law Enforcement
- Currently, it takes a team of two very experienced modelers about two weeks to populate the model and vet it with SMEs. Data need includes that on actors, groups, international organizations, nongovernmental organizations, neighborhoods, and belief systems.
- Athena is a deterministic computer-generated equilibrium type model, focused on belief systems of the different study populations and geographic areas, the relationships between actors, and the popular moods
- Tool frequently used for training with fictitious countries.

ATLAS.ti
- Overview[4]
  - – Center: NWC
  - – Tool/approach: ATLAS.ti
  - – Description: Toolset to manage, extract, compare, explore, and reassemble meaningful pieces from large amounts of data in flexible yet systematic ways
  - – Government sponsor: N/A; COTS product
  - – Developer: Scientific Software Development GmbH
- Information
  - – Focuses: Knowledge management, data collection, analysis, and visualization
  - – Operational levels: All

---

[4]   See ATLAS.ti, homepage, undated.

- Purpose: Qualitative/thematic analysis
- Forces: All
- Current uses: Navy training and Navy wargaming support
- Limitations: Potential permissions and licensing issues for multiple networks
- Requirements
  - Facilities: Minimal facility requirements; workspace for computers only
  - Equipment: Runs on a variety of platforms and operating systems, including Windows
  - Personnel: Individuals trained on ATLAS.ti
  - Required training: Extensive training support offered by vendor, which includes web conferencing; self-paced, face-to-face seminars; and on-site workshops
  - Cost: For government use, a five-user license is $1,950 per year to lease (with free upgrades during the term) or $4,650 to purchase outright (with upgrades purchased separately)

Notes:
- ATLAS.ti employs coding, or electronically marking the data by word, phrases or multiple paragraphs, which can then be organized, referenced and cross-referenced. This aids organizing for review, which leads to better support for analysis and reports but also allows analysts to step back and view the data as a whole to find ideas, associations, or insights (inductive analysis).
- Using standard coding references and event relevant coding, multiple events may then be reviewed to find larger themes or trends
- ATLAS.ti is able to serve as a knowledge management, wargame data collection, and visualization platform. It can be used to analyze the way participants communicate with one another, how they conceptualize problems, and how they interpret the scenario.

Board and Card Games
- Overview
  Centers: CASL, CGSC, CNA, NUWC, RAND
  - Tool/approach: Board and card games
  - Description: Several centers developed board and card games intended to introduce abstract concepts like deterrence and cooperation. More complicated games have been used to communicate strategic and operational concepts.
  - Government sponsors: NDU component colleges, RAND Center for Middle East Public Policy, TRAC
  - Developers: CASL, CNA, RAND
- Information
  - Focus: Any
  - Operational levels: Operational and strategic
  - Purpose: Design

- Forces: Any
- Current uses: NDU uses a board game on COIN, RAND uses an ISIL game for educational purposes, and TRAC works on irregular warfare. The CGSC uses a game to teach joint targeting. Fleet Battle School is used by NUWC to look at aspects of high tactical/low operational–level maritime operations, and Low Resolution Tactical Simulator (Submarine Operations) is used by NUWC and the Undersea Warfighting Development Center to experiment with tactics and concept development.
- Limitations: Limited tolerance for complexity makes these tools better for teaching abstract patterns rather than grappling deeply with a specific case
- Requirements
  - Facilities: Single room with table
  - Equipment: Board and card production; ideally, plotter printer and three-dimensional printer
  - Personnel: Designer with exposure to games of the types; ideally, with past experience as a designer or a lot of time for playtesting
  - Required training: None

Notes:
- A good rule of thumb mentioned by one center is that "serious" board games need to be an "order of magnitude" simpler than most commercial games in order to accommodate lack of time and specialized knowledge[5]
- A small cadre of experts in the play of a particular game can lead a team of players who discuss "what they want to do," and the experienced player can quickly perform a representative move in the game, generating feedback the entire team can take back to the next turns deliberations.

The Caffrey Triangle
Figure B.1 illustrates the Caffrey Triangle.

- Overview
  - Centers: AFMC, J-8 SAGD
  - Tool/approach: The Caffrey Triangle
  - Description: Framework for considering the purpose of a red cell in a game. Depending on where ones purpose falls, one will want to task the red team based on particular constraints mentions under each corner. Hybrids are possible, but may present uncomfortable trade-offs.
  - Government sponsors: AFMC, J-8 SAGD

---

[5] Interview with CNA staff, Arlington, Va., January 4, 2016.

Figure B.1
The Caffrey Triangle

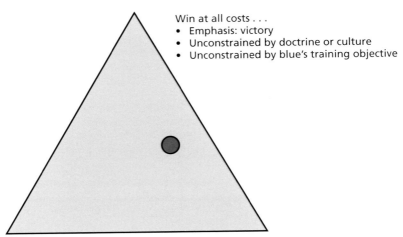

Win at all costs . . .
- Emphasis: victory
- Unconstrained by doctrine or culture
- Unconstrained by blue's training objective

Train me . . .
- Emphasis: a foil for blue
- Unconstrained by doctrine or culture
- Constrained by blue's training objective

Act like "them" . . .
- Emphasis: realism
- Constrained by doctrine or culture
- Unconstrained by blue's training objective

SOURCE: J-8 SAGD.

- Developer: Col Matthew B. Caffrey, Jr. (Ret.), AFMC
- Information
  - Focuses: All
  - Operational levels: All
  - Purpose: Game design
  - Forces: All
  - Current uses: Shaped game design decisions in two (or more) sided games by helping the design team think through the purpose and subsequent design choices
  - Limitations: Framework helps designer understand what setup would be best for a particular purpose, and can help think through the potential costs of shortcoming in the red team, but it is still up to the designer to recruit the right team and provide the correct guidance
- Requirements
  - Facilities: Minimal
  - Equipment: Minimal
  - Personnel: Minimal
  - Required training: None

Close Action Environment

- Overview[6]
  - Center: Dstl
  - Tool/approach: CAEn
  - Description: Computer-based model of urban combat environments that can be used to program the game environment, play interactive games, and then perform stochastic analysis of the results. Best used when detailed and physics-based data on tactical engagements are required.
  - Government sponsor: Dstl
  - Developer: Dstl
- Information
  - Focus: Urban
  - Operational level: Tactical
  - Purposes: Visualization, adjudication, and analysis
  - Forces: Ground
  - Current uses: Human-in-the-loop gaming of urban tactical engagements
  - Limitations: Low-quality graphics make this best suited as an analytical rather than educational tool. Tool takes a considerable time to use, does not handle large forces or environments.
- Requirements
  - Facilities: Minimal
  - Equipment: Speed of runs is dependent on the processors available
  - Personnel: Required trained personnel to set up and supervise
  - Required training: Currently no training program

Notes:

- CAEn is a multisided, computerized wargame and simulation that is capable of representing all-arms close-combat battle from individual soldier or platform to the company or group level. CAEn can model both highly detailed rural and urban terrains in a variety of weather and lighting conditions.
- CAEn is most commonly run as a human-in-the-loop, closed wargame, with commanders only knowing what their units have detected. Output is qualitative (e.g., insights supported by data) and quantitative (e.g., analysis of engagements).
- CAEn is primarily used to investigate the impact of changes on the scheme of maneuvers, capability, and force structure on the ability of company and below forces to complete the mission.

Combat Operations in Denied Environments

- Overview
  - Center: RAND

---

[6] See Mark Gould, *The CAEn Process—Wargaming, Simulation and Replication*, London: UK Ministry of Defence, Defence Science and Technology Laboratory, September 8, 2016.

- – Tool/approach: Combat Operations in Denied Environments
- – Description: A suite of four models that identifies detailed lists of combat support requirements for various basing postures and scenarios, optimal locations for war reserve material, and investments in defense resources and their effect on the Air Force's ability to conduct operations in denied environments
- – Government sponsor: Air Force
- – Developer: RAND
- Information
  - – Focus: M&S trade-offs among various basing strategies
  - – Operational levels: Theater and strategic
  - – Purpose: Force planning. This aggregation of models determines manpower and equipment requirements for a given basing posture in a combat scenario; the optimal location to store and maintain support resources for rapid deployment; how various infrastructure investments, defense options, recovery capabilities, and CONOPS perform under simulated missile attack; and the most cost-effective means to improve sortie generation. Overall, it allows decisionmakers a robust exploration of trade-offs among different basing strategies and resource investments.
  - – Force: Air
  - – Current uses: Decision support for Air Force in determining the required resources for various basing strategies
  - – Limitations: May require access to restricted databases that are used as inputs into some of the models. Many model factors are constrained to an Air Force context. Expanding some of these factors, such as fuel, electricity, materiel delivery, and munitions, will require joint input since they often involve interservice coordination.
- Requirements
  - – Facilities: Minimal
  - – Equipment: Computers able to run Excel, the General Algebraic Modeling System, and Java applications
  - – Personnel: Usable by a single individual
  - – Required training: Models are accessible through relatively straightforward GUIs and configuration files and require minimal training

Command: Modern Air Naval Operations

- Overview[7]
  - – Tool/approach: *Command: Modern Air Naval Operations*
  - – Description: Commercial computer game with increasingly professional pedigree involves a single-user "sandbox" to create and walk through maritime warfare scenarios in real time to approximately ten times the speed of real

---

[7]   See WarfareSims.com, "Command: Modern Air/Naval Operations," webpage, undated a.

time. Windows PC software application with extensive open-source database of ships, submarines, aircraft, spacecraft, and related ground installations and defenses. Scenarios can be rapidly created from an extensive library of prebuilt platforms and installations, modified and rerun.
- Government sponsor: N/A; COTS product
- Developer: WarfareSims.com[8]
- Information
  - Focuses: All
  - Operational levels: Tactical to low operational
  - Purpose: Scheme of maneuver exploration and visualization
  - Forces: Naval, air, and space; limited ground (static installations and defenses)
  - Current uses: Stand-alone exploratory analysis tool to understand broad synergies across platforms and related capabilities/limitations. Visualization of the dynamics of a set "scheme of maneuver" across time/space/force dimensions.
  - Limitations: Currently allows for only a single player versus a "programmed opponent" plan. One side follows a set plan, while the analyst exploring the scenario can dynamically play the other side.
  - A professional version is available with additional capabilities, but at a substantial additional cost
- Requirements
  - Facilities: Suitable for any office environment
  - Equipment: Windows PC with moderate graphics capability
  - Personnel: None; stand-alone software application
  - Required training: Developing familiarity with the game requires about two weeks of dedicated "stick time" playing the game. It is about as complicated as commercial computer wargames get. Training is available from WarfareSims to speed this timeline, but at a cost.

Configurable Physical Gaming Space
- Overview
  - Centers: CNA, CSL, DST, Dstl, NWC
  - Tool/approach: Configurable physical gaming space
  - Description: Large, configurable space for gaming. Ideally, the space can be subdivided into spaces of different sizes, have furniture that can be easily moved, and allow for custom AV and IT configurations.
  - Government sponsors: DST, Dstl, NWC
  - Developers: DST, Dstl, NWC
- Information
  - Focuses: All
  - Operational level: Any

---

8   WarfareSims.com, homepage, undated b.

- Purpose: COP
- Forces: All
- Current uses: Diverse games, including computer-assisted games, manual board games, and matrix- and seminar-style games
- Limitations: Initial outlay can be more expensive, but the flexibility it provides far outweighs the cost. Fixed configuration spaces may be more "flashy," as they may provide cool visuals, yet the ability to rearrange spaces to fit game design versus designing the game around the furniture provides game designers and developers much more latitude. Historically, the WGD has found that spaces with fixed furniture or single-use design actually get used less than spaces that are able to be reconfigured.
- Requirements
  - Facilities: Large, dedicated physical space with movable partitions, furnishings, and tech equipment (including AV and network infrastructure)
  - Equipment: AV equipment to support production and VTC. Enabling communication between control and teams, and between teams, is critical.
  - Personnel: Facilities and AV support
  - Required training: As needed to augment facilities and AV infrastructure

The Cycle of Research

Figure B.2 illustrates the Cycle of Research.

- Overview
  - Center: CNA
  - Tool/approach: The Cycle of Research
  - Description: A framework for how wargames fit with analysis and live exercises into a continuous cycle of research
  - Government sponsor: N/A
  - Developer: Peter Perla
- Information
  - Focuses: All
  - Operational levels: All
  - Purpose: Framework for integrating wargames with analysis and live exercises
  - Forces: All
  - Current uses: CNA, NUWC
  - Limitations: Relevant for situations where there are wargames, live exercises, and analytic efforts to integrate
- Requirements
  - Facilities: None
  - Equipment: None
  - Personnel: Minimal
  - Required training: Understanding of research

Figure B.2
The Cycle of Research

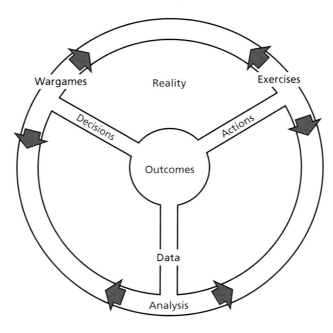

SOURCE: Perla, 2012.

The Day After Methodology
- Overview
  - Center: RAND
  - Tool/approach: The Day After methodology
  - Description: An exploratory gaming methodology originally developed following the collapse of the Soviet Union to study the range of issues surrounding nuclear proliferation. Utilizes a series of multistage case study exercises designed to focus experts and decisionmakers on the concrete issues of a particular policy dilemma. Groups of participants were given a crisis to which they needed to devise an effective policy response. Each scenario begins on "the day of" a significant policy crisis. Participants must then devise a policy response to a significant escalation "the day after." Finally, players step back and, in light of exercise events, consider the range of prospective actions, policies, and plans that could have been taken "the day before" and might have prevented the crisis.
  - Government sponsor: Air Force
  - Developer: RAND
- Information
  - Focuses: Exploratory analysis, stakeholder engagement, and education
  - Operational level: Strategic planning

- Purpose: To identify salient issues and better understand the various policies and strategies that might be considered by top decisionmakers during a crisis event. If the same scenario is run multiple times with different groups, it is possible to aggregate "patterns of response" that can then feed into more elaborate wargames.
- Forces: High-level decisionmakers
- Past Uses: Nuclear proliferation, money laundering and financial technology, and information warfare
- Current uses: Critical infrastructure and disaster scenarios for the Department of Homeland Security, public health emergencies for local and state governments, and exploring the utility of a new technology
- Limitations: Requires expert scenario development to be effective. Works best for crisis and discontinuity events that are of low probability but high consequence.
- Requirements
  - Facilities: Minimal facility requirements. Requires space for groups to meet and collaborate (ideal size is three groups of eight individuals to each work on the same scenario).
  - Equipment: Minimal equipment. A means of communicating the scenario: PowerPoint, projectors, and the like.
  - Personnel: Skilled exercise designers and scenario writers, a skilled facilitator working with each group to enforce scenario and agenda. Note takers to capture discussions.
  - Required training: No training for participants. Skilled facilitators can provide guidance during gameplay, but there is no rule set and constraints are minimal.

Notes:
- The testing of scenarios makes this exercise expensive. One must create a plausible future history, its context, and the boundary conditions of the problem the client wants to explore. Afterward, one must perform test runs with SMEs. Then one must tweak and modify as necessary. The exercise requires at least two test runs with in-house SMEs, then another test with outside SMEs. This process ensures validity and accuracy of the scenario. Then one should test the exercise with the staff of the senior leadership before executing the operational exercise with actual leaders. Testing it with the staff ensures the exercises playability. Developing good scenarios can take a year, and this process might need to be curtailed if the timeline is only six to seven months.
- One advantage of this methodology is the playing time. If one wants to have high-level participants, one wants a game that can last no more than three to four hours, which is perfectly possible with this type of game. Games can be stretched out over two days for larger multilateral exercises, but with an ideal group size, they can be completed in a morning or afternoon.

Decision Lens

- Overview[9]
  - Center: J-8 SAGD
  - Tool/approach: Decision Lens
  - Description: Decision Lens is an end-to-end software solution and process for identifying and prioritizing, analyzing, and measuring which investments, projects, or resources will deliver the highest returns. It allows organizations to immediately see the impact and trade-offs of the choices they make.
  - Government sponsor: N/A; COTS product
  - Developer: Decision Lens
- Information
  - Focuses: Data analytics and planning
  - Operational levels: Operational and strategic
  - Purpose: To reduce risk and improve outcomes by facilitating collaboration. Allows stakeholders to more easily discuss key trade-offs and allocation decisions through scenario-based planning.
  - Forces: All
  - Current uses: J-8 SAGD has used Decision Lens in its games and tabletop exercises to help a broad group of people prioritize on a set of criteria. Participants will be given a single criterion and asked to vote anonymously for the option that better meets that criterion. The results of these quick surveys will be used to generate discussion and form consensus on key definitions and considerations.
  - Limitations: Commercial software can be difficult to use on classified networks. Users have expressed some difficulty in getting the appropriate licenses and approval.
- Requirements
  - Facilities: Conference room with projector
  - Equipment: Suitable desktop computer to run software and hardware for interactive voting system
  - Personnel: N/A
  - Required training: N/A

Decisive Action

- Overview
  - Center: CGSC
  - Tool/approach: Decisive Action
  - Description: Computer-based tool modeling modern divisional- and corps-level combat and decisionmaking

---

[9] See Decision Lens, homepage, undated.

- Government sponsor: CGSC
    - Developer: Decisive Point LLC
- Information
    - Focuses: Decisionmaking and staff exercise driver
    - Operational level: Tactical
    - Purpose: Education
    - Forces: Ground
    - Current uses: Classroom support at the CGSC
    - Limitations: Designed to illustrate a variety of high-intensity conflict situations focused on land warfare
- Requirements
    - Facilities: Minimal
    - Equipment: Laptop computer
    - Personnel: Few
    - Required training: Operator training is four hours

Dilemma Analysis
- Overview
    - Center: Dstl
    - Tool/approach: Dilemma analysis (also called conformation analysis)
    - Description: Method to structure multiparty conflicts (including negotiations and more active conflicts) by collecting information on priorities, statements from actors about their preferences, the potential results of not resolving the dispute, and where distrust exists between players. Currently, the software can help highlight areas for discussion, but the process is likely helpful just as a tool to structure discussion.
    - Government sponsor: Dstl
    - Developer: Dstl
- Information
    - Focus: Any
    - Operational level: Any
    - Purpose: Game design and adjudication
    - Forces: Ground
    - Current uses: Current experiments use this in workshops to help shape game design
    - Limitations: Not a game per se
- Requirements
    - Facilities: Minimal
    - Equipment: Minimal
    - Personnel: Requires trained personnel to set up and supervise
    - Required training: Currently no training program

Design Thinking
- Overview
  - Center: SOCOM
  - Tool/approach: Design thinking
  - Description: Design thinking is a process that combines in-depth constituent or consumer insights and rapid prototyping to generate innovative, effective solutions. Inherently experimental, design thinking seeks to better understand, emphasize, and solve the needs of its target audience.
  - Government sponsor: N/A
  - Developer: IDEO
- Information
  - Focus: Any
  - Operational level: Any
  - Purpose: Game design and facilitation
  - Forces: Any
  - Current uses: Used to cultivate innovative thinking and problem-solving skills among action officers. SOCOM often employs design thinking in conjunction with SSM for its wargaming program. Using design thinking, facilitators craft techniques and processes to illicit novel and useful ideas from participants.
  - Limitations: Not specifically designed for wargame design, and its utility largely depends on the proficiency of the designer
- Requirements
  - Facilities: N/A
  - Equipment: N/A
  - Personnel: Minimal
  - Required training: Individuals can attend courses offered by the Joint Special Operations University, which can be seven to ten days long. Civilian institutions like Stanford University also offer more intensive courses on design thinking.

Notes:
- The Joint Special Operations University offers seven- to ten-day courses on design thinking. The classes aim to encourage and build critical thinking skills for action officers within SOCOM. Classes include the following:
  - Introduction to Design Thinking
  - Design Thinking for Practitioners
- Stanford offers online resources and a virtual online course on design thinking[10]

---

[10] See Stanford University, d.school, "Tools for Taking Action," webpage, undated.

- Hacking for Defense, a Stanford course, teaches students and government sponsors to utilize design thinking and the Lean Launchpad methodology to collaborate and generate innovative solutions to real-world problems[11]
- Harvard University also offers a similar course on design thinking, Creative Thinking: Innovative Solutions to Complex Challenges.[12]

Dynamic Morphological Exploration Trees
- Overview
  - Center: DST
  - Tool/approach: DME trees
  - Description: An add-on approach developed by DST to GMA that creates a tree mapping of optimal search paths through a morphological space
  - Government sponsor: DST
  - Developer: DST
- Information
  - Focus: Any
  - Operational levels: Tactical and operational
  - Purpose: Analysis
  - Forces: Any
  - Current uses: Examination of combat vehicle variations
  - Limitations: Requires a defined criterion for success in order to score different branches of the DME tree. This identifies the most promising cases for human-in-the-loop wargaming, but requires a considerable amount of time.
- Requirements
  - Facilities: Minimal
  - Equipment: Computer tool to assist in instantiation of the tree
  - Personnel: OR or similar technical analysts
  - Required training: Training in GMA, some level of programming skills likely helpful

Notes:
- Used within a process that combines GMA, DME trees, and wargaming:
  - Step 1: Generate GMA space
  - Step 2: Use DME trees to identify most promising points or configurations in the GMA space
  - Step 3: Wargame the points in the GMA space identified through DME.

---

[11] See Stanford University, "Hacking for Defense: Class Details," webpage, undated.

[12] See Harvard Extension School, "Creative Thinking: Innovative Solutions to Complex Challenges," webpage, undated.

- For use with GMA, where the points in the GMA space have a criteria by which one can rate "success" for each. Not compatible with all GMA fields—it will depend on how the field was defined.
- DME trees do not overcome the issue of identifying an initial set of points in the GMA space from which to proceed
- Appears more consistent with engineering configuration applications of GMA rather than an application such as managing a future scenario space
- Demonstrated, alternative means of sampling from a GMA space include
  - "interesting corners," where experts defining the field identify unusual combinations of GMA parameters
  - Latin hypercube
  - scenario diversity analysis.
- The Australian Society for Operations Research has additional information on DME trees.[13]

Enterprise Metacard Builder Resource
- Overview
  - Center: M&SCO
  - Tool/approach: Enterprise Metacard Builder Resource and Enterprise Metacard Builder Resource 2.0
  - Description: Knowledge management system based on Extensible Markup Language metadata; focuses on people, resources, and organizations for M&SCO, but the forms are customizable. Tool can stand alone or can interact with the M&S catalog. May be used to create, modify, and publish records for enterprise discovery. May be Common Access Card–enabled. Forms are customizable, and information exportable to CSV files.
  - Government sponsor: M&SCO
  - Developer: SimVentions, but software is GOTS
- Information
  - Focus: Any
  - Operational level: Any
  - Purposes: Data collection, data visualization, knowledge management
  - Forces: Any
  - Current uses: M&S catalog knowledge management system to manage records (at embr.msco.mil)
  - Limitations: No history of use in wargaming
- Requirements
  - Facilities: None

---

[13] Peter B. Williams and Fred D. J. Bowden, "Dynamic Morphological Exploration," paper presented at the 22nd National Conference of the Australian Society for Operations Research, Adelaide, Australia, December 1–6, 2013.

- Equipment: Unclassified cloud implementation on cloud services (SimVentions maintains on cloud, security patches, etc.). Classified instance on the Defense Information Systems Agency cloud, or on classified server.
- Personnel: SimVentions personnel to customize the Enterprise Metacard Builder Resource for Marine Corps wargaming and manage updates; would take four to six weeks for adaptations from the M&S catalog, up to six to nine months for a very complex customization with multiple databases. Marine Corps staff to maintain once developed, but shifting away from maintaining SharePoint sites.
- Required training: Training and user help guides; users able to use within an hour with training. Training and support available on the web and in person.

Notes:
- Potential knowledge management tool for wargaming.

FacilitatePro
- Overview
  - Center: NWC
  - Tool/approach: FacilitatePro
  - Description: FacilitatePro provides brainstorming, prioritizing, evaluating, surveying, and action planning tools to aid creativity and solve complex problems. FacilitatePro is used in meeting rooms to improve group productivity.
  - Government sponsor: N/A; COTS product
  - Developer: Facilitate.com[14]
- Information
  - Focus: Knowledge management
  - Operational levels: All
  - Purposes: Collaboration and communication
  - Forces: Naval
  - Current uses: Navy training, Navy wargaming support
  - Limitations: Licensing issues do exist, but the developer is willing to work with customers to find a cost-effective solution that will provide adequate licensing while recognizing that not all licenses will be in use at one time. (Note: The NWC wishes to emphasize that the technical support is excellent.)
- Requirements
  - Facilities: Minimal facility requirements; requires computer terminals with browser access
  - Equipment: PC running Windows 2003/2008 Server/Vista/Windows 7; 2 gigabytes of free disk space; 2 gigahertz processor or faster; 512 megabytes of available random-access memory

---

[14] Facilitate.com, homepage, undated.

    – Personnel: Individuals trained in FacilitatePro
    – Required training: Two days of basic training included with annual license
    – Cost: License is $19,740 for first year, $4,995 for subsequent years. Includes two-day training for up to eight facilitators, as well as all release and version upgrades.

Fleet Battle School

- Overview
  - Prototype distributed wargaming concept
  - Tool/approach: Fleet Battle School
  - Description: A multilevel command and control architecture for multiplayer maritime wargaming. Multiplatform (Windows/MAC/Linux/Android/iOS) software tools for high tactical/low operational maritime wargaming. A PC is used to create the game using editors for the platform capabilities, the composition of each sides forces, the map, and the combat tables. Each player is given a tablet to give commands to his or her respective forces (drawing the path and giving orders and the guidance to "draw with your finger" on the device). A player only gets the situational awareness on his or her tablet appropriate to the force (which, if it is a single submarine, could be very little). The overall commander accepts the orders, and they may be viewed collectively on a screen driven by the PC. The game will then step through the adjudication process, either based on dice rolls and combat tables, or the umpire can at any point step in and "override the dice" where necessary/appropriate/interesting to the capabilities under consideration.
  - Government sponsor: NUWC
  - Developer: John Tiller Software[15]
- Information
  - Focuses: All
  - Operational levels: High tactical to low operational (units are naval task groups, air squadrons, individual submarines, and shore facilities)
  - Purpose: Exploring the impact of capability changes on scheme of maneuver and CONOPs
  - Forces: Naval, air, space, and limited land forces (static installations and defenses)
  - Current uses: Exploration of the effect of changing the capabilities of one side or both sides platforms over the course of several games to look at where a tipping point occurs and a new set of CONOPS or dramatically different scheme of maneuver is developed. Used by NUWC in several exploratory studies to

---

[15] John Tiller Software, homepage, undated.

look at theater antisubmarine warfare CONOPS. Also for team building/general warfare area familiarity.

- Limitations: Currently in a prototype stage (postalpha), so still in need of considerable playtesting and development. Still effectively a "proof of concept" demonstrator.

- Requirements
  - Facilities: A large room where players can go off to input their orders, or multiple rooms where teams are permanently separated during the game
  - Equipment: Windows PC with moderate graphics capability, a tablet for each player or subcommand (a single tablet can be shared by a group)
  - Personnel: One or more umpires, while the number of players depends on the number of independent commands the analysis calls for
  - Required training: Players can be up and playing the game in about 20–30 minutes by playing a few of the simple tutorial scenarios
  - Tutorial demonstrator is currently available in the Android and Apple app stores

Future Analytical Science and Technology

- Overview
  - Center: AFRL
  - Tool/approach: FAST
  - Description: Series of vignettes based on Title 10 game scenario that enables players to pick and utilize a wide range of technology in proof-of-concept phase of development to better understand impact on the battlefield
  - Government sponsor: Air Force, as part of lead-up to Title 10 game series
  - Developer: AFRL

- Information
  - Focuses: Technology and capability assessment
  - Operational levels: Operational and tactical
  - Purpose: Game Design and Analysis
  - Forces: All, but air focus
  - Current uses: Air Force Title 10 inputs
  - Limitations: Not intended to validate performance, but rather to test whether and how new technology can be useful in solving tactical and operational problems (largely qualitative output)

- Requirements
  - Facilities: Space for games; currently tending to use the gaming center at Maxwell Air Force Base, as it comes with free staff support
  - Equipment: Minimal
  - Personnel: Five people over two months to develop, five people to execute (might be more at a different venue)
  - Required training: Unclear

Notes:
- FAST allows for quick examinations of plausible technologies before serious study of feasibility and broader buy-in occurs as part of the Title 10 process
- FAST uses vignettes drawn from the broader Title 10 Global Engagement scenario, as they are already vetted and known to be relevant. Requirement for new scenario would add to the game preparation time and staff required.

Gaming Auditoriums
- Overview
  - Center: SOCOM
  - Tool/approach: Gaming auditoriums
  - Description: Dedicated space for gaming in the form of a large, auditorium-style gaming room consisting of a large map surrounded by auditorium seating. Supported by extensive AV equipment to allow for presentation and VTC connectivity.
  - Government sponsor: SOCOM
  - Developer: SOCOM
- Information
  - Focuses: All
  - Operational level: Any
  - Purpose: COP
  - Forces: All
  - Current uses: Senior leader seminar and MAPEX-style games
  - Limitations: Lack of flexibility makes it difficult to repurpose space for smaller groups; tends to work best with seminar MAPEX games, but can limit collaboration for matrix- or board game–style designs. Also limits multicell play.
- Requirements
  - Facilities: Large dedicated physical space, which can be repurposed for VTC and briefings
  - Equipment: AV equipment to support production and VTC
  - Personnel: AV support
  - Required training: As needed to augment AV infrastructure

Google Drive and Google Sites
- Overview
  - Center: NWC
  - Tool/approach: Google Drive and Google Sites
  - Description: Google Sites includes a wide range of applications such as chat, docs, calendar, maps, talk, and more. Specifically, Google Drive allows for the creation of websites that act as a secure place to store, organize, share, and access information. Similar to DropBox and TeamSpace, Google Drive allows

for both storage and collaborative work through a cloud-based information system.
- Government sponsor: N/A
- Developer: Google
- Information
    - Focuses: Mapping, visualization, knowledge management, communication, and cooperation
    - Operational level: N/A
    - Purpose: Used for information sharing and knowledge management during wargames
    - Forces: N/A
    - Current uses: The NWC integrates Google Drive and Google Sites for knowledge management during wargames and collaborative work
    - Limitations: Limited extensibility, since the software is not government owned or contracted
- Requirements
    - Facilities: PC and internet connection
    - Equipment: Runs on Linux, Mac OSX, and Windows
    - Personnel: Depends on IT policies
    - Required training: Minimal training
    - Cost: Free

Google Earth
- Overview
    - Centers: CAA, NWC
    - Tool/approach: Google Earth
    - Description: Google Earth is a virtual globe and mapping tool that allows viewing of satellite imagery, maps, and terrain. NWC uses both the online Google Earth on the UNCLASS gaming network, and has purchased Google Earth Enterprise licenses for classified gaming networks. These licenses are no longer available, as the product is being deprecated and will no longer be supported by Google after June 2017.
    - Government sponsor: N/A
    - Developer: Keyhole, Inc., purchased by Google in 2004
- Information
    - Focuses: Mapping and visualization
    - Operational levels: All
    - Purpose: Can be integrated into a wargaming center's COP for use by players, used during presentations as a visual aid, or as a source of printed high-resolution satellite imagery for game boards or other visual aids
    - Forces: Naval, ground, and air

- Current uses: NWC integrates Google Earth imagery into some of its tools to increase immersion and situational awareness for players
- Limitations: Limited extensibility, since the software is not government owned or contracted
- Requirements
    - Facilities: Depends on use; Google Earth Enterprise requires a dedicated Red Hat Enterprise Linux server
    - Equipment: Runs on Linux, Mac OSX, and Windows
    - Personnel: Programming support for integration into current software tools
    - Required training: Minimal training to use as a player or for presentations Annotating maps requires knowledge of Keyhole Markup Language
    - Cost: Free

The Grand Offensive
- Overview
    - Center: CGSC
    - Tool/approach: *The Grand Offensive*
    - Description: Game to drive history class on World War I. Students plan the offensive on the Somme
    - Government sponsor: CGSC
    - Developer: CGSC
- Information
    - Focus: Decisionmaking
    - Operational level: Tactical
    - Purpose: Education
    - Forces: Ground
    - Current uses: Classroom support at the CGSC
    - Limitations: Designed to illustrate problems faced by commanders in World War I
- Requirements
    - Facilities: Minimal
    - Equipment: Mac, Windows, or web access
    - Personnel: None
    - Required training: None

Green Country Model
- Overview
    - Center: JHU-APL National Security Analysis Department
    - Tool/approach: GCM
    - Description: A wargame that simulates civilian (green) effects in a red-versus-blue conflict. JHU-APL designed GCM in order to model the social and societal factors at play during unconventional, irregular warfare and stabil-

ity operations. Unlike large-scale conventional warfare, these operations can hinge on the relative influence and perceived legitimacy of state and non-state actors by a civilian population. GCM is adapted for human-in-the-loop gameplay.

- Government sponsor: Various, including the AFRL, Northern Command, and the Office of Naval Research
- Developer: JHU-APL

• Information
- Focus: Sociocultural M&S
- Operational levels: All
- Purpose: To better understand and simulate the Diplomatic, Intelligence, Military, Civil Affairs, and Economic parameters at play in a real war. By simulating nonplayer characters in a wargame between red and blue teams, players can better understand the range of "soft" factors that real warfare often hinges on. While more traditional simulations force a war of attrition, GCM attempts to incorporate a wider range of factors that provide options for collaboration and cooperation, as well as more traditional kinetic conflict.
- Forces: Ground, and nonmilitary organizations such as nongovernmental organizations, local governments, religious groups, or crime syndicates
- Current uses: Used in games to represent and understand regions of Central America, Nigeria's role in oil production and export, and China's emerging economic and commodity market
- Limitations: Requires the input of SMEs to accurately describe and calibrate the model for desired nonplayer characters. Only two sides (red and blue) are playable. GCM is a high-level, stochastic, multisided competitive influence game that excels at modeling interagency dynamics. Conventional force-on-force conflicts (and their attendant detailed models of weapon systems and sensors) are not suitable in GCM.

• Requirements
- Facilities: Minimal facility requirements. A screen on which to project a game map for all players to see, and space for all participants.
- Equipment: GCM is based on a series of Excel-based models
- Personnel: Moderators and adjudicators during gameplay. Teams of players control red and blue forces while nonplayer characters are controlled by the simulation and are programmed prior to the game. SMEs to program realistic scenarios and NPA behavior.
- Required training: Not specified

Harpoon
• Overview
- Center: CNA
- Tool/approach: Harpoon

- Description: Harpoon is a commercial board game and computer game for two to eight players that covers various aspects of maritime combat, including surface, subsurface, and air engagements. It contains a detailed and comprehensive rule set covering many aspects of modern naval engagement.
    - Government sponsor: N/A
    - Developer: Adventure Games/Clash of Arms Games
- Information
    - Focus: Game design
    - Operational levels: Tactical and operational
    - Purpose: Up-to-date weapon and platform ratings allow for relatively realistic evaluations of scenarios
    - Forces: Naval
    - Current uses: Used, along with other classic board games, as a potential source of game mechanics and inspiration during the design phase
    - Limitations: N/A
- Requirements
    - Facilities: N/A
    - Equipment: Table for board game or PC for computer game
    - Personnel: Two to eight players
    - Required training: None

i2 Analyst's Notebook
- Overview
    - Center: NWC
    - Tool/approach: i2 Analyst's Notebook
    - Description: i2 Analyst's Notebook allows users to rapidly piece together disparate data into a single cohesive picture; identify key people, events, connections, and patterns; increase understanding of the structure and hierarchy of data; and generate visualizations
    - Government sponsor: N/A; COTS product
    - Developer: IBM
- Information
    - Focus: Knowledge management
    - Operational level: Any
    - Purpose: One major part of "analysis up front" is structuring the collected data from an event with the subsequent analysis needs in mind. The challenge of qualitative analysis typically amounts to addressing the volume of data. However, grouping and associating the data reduces the volume, which allows analysts to quickly collate, analyze, and visualize data from disparate sources. Moreover, it reduces the time required to discover key information in complex data.
    - Forces: Any

- Current uses: Qualitative analysis, Navy wargaming support
- Limitations: Potential permissions and licensing issues for multiple networks
- Requirements
  - Facilities: Enterprise-level database installed; analysts require direct access
  - Equipment: Minimum of 1 gigabyte free disk space for installation. Total disk space requirements depend on the planned size of the repository; an average local analysis repository will use 150 gigabytes of space. Minimum of 4 gigabytes random-access memory; 2 gigahertz quad-core processor. Software requirements include Apache Derby, Java SDK, Microsoft .NET Framework, Visual C++, and a web browser.
  - Personnel: Individuals trained in i2 Analyst's Notebook
  - Required training: Three-day instructor-led course
  - Cost: $7,000–$19,000 per license

Notes:
- Given the increased volume of electronically collected data, a graphic means to usefully and usably approach and display trends or outliers helps relay important factors quickly and meaningfully. In 2009 the NWC Wargaming Department procured i2 Analyst's Notebook, a data visualization tool widely used in the intelligence community to identity connections between relational databases.
- Utilizing the structure of the data, associations and occurrences are displayed in several graphic formats of link and node, theme line for progressive display or timeline, or even clusters and groups to convey some aspect supporting or refuting other results. The challenge of qualitative analysis typically amounts to addressing the volume of data. However, grouping and associating the data reduces the volume, and this allows analysts to quickly collate, analyze, and visualize data from disparate sources. Moreover, it reduces the time required to discover key information in complex data. Graphic chart output in PDF format can be directly included in reports.[16]

i2 iBridge and i2 iBridge Designer
- Overview
  - Center: NWC
  - Tool/approach: i2 iBridge and i2 iBridge Designer
  - Description: i2 iBridge is an advanced connectivity and analytical search solution that connects i2 Analyst's Notebook users directly to enterprise databases. iBridge Designer allows analysts to craft specific queries outside i2 Analyst's Notebook to extract data from the database.
  - Government sponsor: N/A; COTS product
  - Developer: IBM

---

[16] See IBM, "IBM i2 Analyst's Notebook," webpage, undated.

- Information
  - Focus: Knowledge management
  - Operational level: Any
  - Purpose: Provides analysts with immediate access to search and query options directly from the i2 Analyst's Notebook database task pane. Connecting directly to the information stored in the database, analysts can see changes as they occur and can respond immediately, providing quicker analysis and more efficient adjudication of moves.
  - Forces: Any
  - Current uses: Qualitative analysis, Navy wargaming support
  - Limitations: Users must have intimate knowledge of the database and structure
- Requirements
  - Facilities: Enterprise-level database installed; analysts require direct access
  - Equipment: Hardware minimums similar to i2 Analyst's Notebook
  - Personnel: Individuals trained in i2 Analyst's Notebook; basic database knowledge (tables, views, relations, and keys) is necessary
  - Required training: One-day instructor-led course
  - Cost: Around $400 per license for i2 iBridge; around $1,200 per license for i2 iBridge Designer

Notes:
  - The NWC Wargaming Department has been using i2 iBridge since 2012 and has used it extensively during major analytical games
  - The NWC maintains one iBridge Designer and three i2 iBridge licenses for use during games
  - i2 iBridge basically functions as a plug-in to i2 Analyst's Notebook, activating a database task pane within the application. As the analyst executes a query, data contained in the database is immediately displayed, providing analysts and white-cell adjudicators insight into player actions.[17]

Interactive Process of Inquiry for Capability Development

- Overview
  - Center: NUWC/Naval Sea Systems Command Fleet Engagement Community of Practice
  - Tool/approach: Interactive process of inquiry for capability development
  - Description: Framework for integrating the activities related to capability development—expanding on the Perla cycle of research to include aspects of the philosopher Charles Sanders Peirce's process of inquiry. Figure B.3 illustrates the interactive process of inquiry.
  - Government sponsor: NUWC
  - Developer: Paul Vebber, NUWC

---

[17] For an example of i2 iBridge Designer in action, see https://www.youtube.com/watch?v=8h9OrYWlHmY.

Figure B.3
The Interactive Process of Inquiry

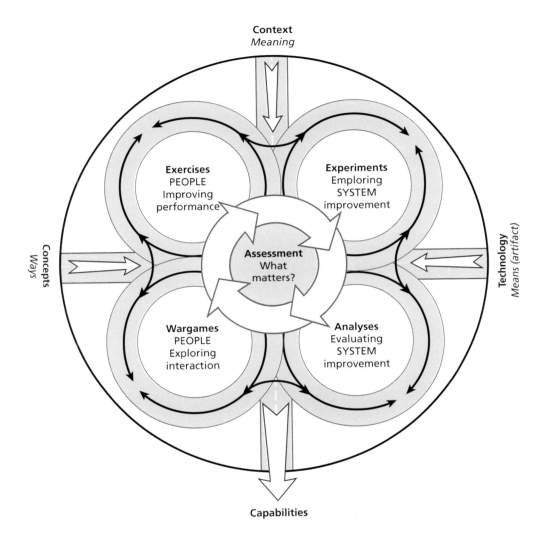

- Information
  - Focuses: All
  - Operational levels: All
  - Purpose: Planning and integration
  - Forces: All
  - Current uses: Used to take activities in the capability development process and organize them into those that involve wargaming (people exploring inter- actions, and decisionmaking-related interactions in particular); analysis (people evaluating improvements to systems); exercises (people improving their

performance); and experiments (exploring how systems might be improved). This is all moderated by an assessment process that looks for "what matters" in the given context. The process has inputs and an output: a context (typically related to some aspect of warfighting), concepts for the ways advantage can be found within the context, and means that are thought to provide advantage. The relationships between the roles of the processes of analysis, wargaming, exercises, and experimentation across the levels of war is indicated. A descriptive framework for the analysis and wargaming activities involved is given in terms of how the forces and decisionmakers are represented and how adjudication is performed, together with how consequential player decisions are—the more consequential, the more gamelike. Where the decisions are not consequential, or are not actually being made, the event is thought of as more of a workshop. Currently being developed as part of a prototype process to support the Fleet Design concept development effort and Fleet Forces Command.

– Limitations: Any process is only as good as the inputs and, in this case, the assessment process that works across the core functions

Figure B.4
Roles and Relationships vs. Levels of War (Analysis, Wargaming, Exercise/Experimentation)

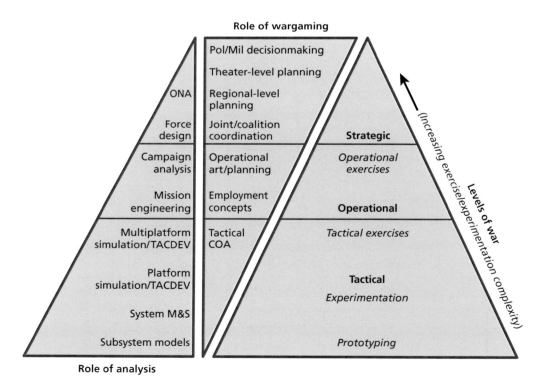

Figure B.4—Continued

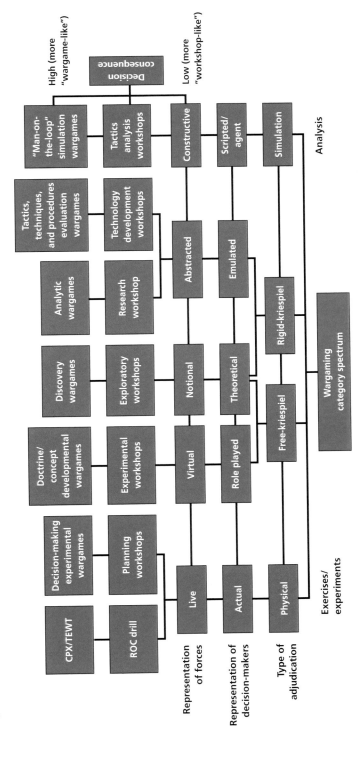

SOURCE: NUWC.

- Requirements
  - Facilities: None
  - Equipment: None
  - Personnel: None
  - Required training: None

Jabber
- Overview
  - Center: SOCOM Wargame Center
  - Tool/approach: Jabber
  - Description: Jabber streamlines communications by integrating instant messaging, video, voice, voice messaging, screen sharing, and conferencing capabilities securely into one client on your desktop
  - Government sponsor: N/A; COTS product
  - Developer: Cisco Systems
- Information
  - Focus: Knowledge management
  - Operational level: Any
  - Purpose: Cisco's Jabber platform comes with a variety of tools to enhance collaboration between game participants and improve data retention for later analysis
  - Forces: Any
  - Current uses: Jabber can be used as a convenient platform to take notes and for ongoing data collection so that all information is consolidated and time-stamped
  - Limitations: Potential permissions and licensing issues for multiple networks
- Requirements
  - Facilities: Minimal
  - Equipment: Desktop computers for data entry
  - Personnel: Potential IT support required for software installation
  - Required training: Minimal

Notes:
  - In September 2008, Cisco Systems acquired Jabber, Inc., the creators of the commercial product Jabber XCP. The original Extensible Messaging and Presence Protocol on which Jabber was based is still an open standard. Cisco hardware and service contracts would be required for some of the other integration and collaboration tools, but free software implementations of the chat service are available.[18]

---

[18] Cisco, "Cisco Jabber," webpage, undated.

Joint Semi-Automated Forces

- Overview
  - Centers: AFRL Human Effectiveness Directorate, Navy Warfare and Development Command, NWC
  - Tool/approach: JSAF
  - Description: JSAF is a computer-generated forces system that provides entity-level simulation of ground, air, and naval forces. It is the primary framework for ongoing research and development in human behavior representation. It is also being used to support a variety of experiments with environment simulation, including dynamic terrain (i.e., craters, trenches), weather, and chemical/biological warfare defense.
  - Government sponsor: The Navy Warfare and Development Command is the current program manager
  - Developer: JSAF was developed as part of the Defense Advanced Research Projects Agency Synthetic Theater of War Advanced Concept Technology Demonstration
- Information
  - Focuses: M&S for training and experimentation, wargame adjudication, visualization, and planning
  - Operational levels: All
  - Purpose: JSAF is a simulation system used to generate entity-level units such as tanks, ships, aircraft, and individual combatants, including their associated sensor systems and munitions. The individual units can be controlled individually or collectively. JSAF units execute tasks and behaviors appropriate to the type of unit, and users may override or interrupt many of the automated behaviors. JSAF is designed to be extensible, customizable, and high-level-architecture compliant to allow interoperability with other simulation systems. It can also plug into real workstations in order to provide more realistic training to service members.
  - Forces: Naval, air, and Joint Forces Command
  - Current uses: Used by the U.S. Joint Forces Command Navy Warfare Development Center for joint experimentation, by the U.S. Navy for Fleet Battle Experiments, and by the AFRL Human Effectiveness Directorate in support of the Distributed Mission Training program. It has been used to simulate up to 40,000 entities within a single distributed simulation. The NWC also uses it for wargame adjudication, visualization, and planning.
  - Limitations: JSAF's complexity may make it unsuitable for smaller, more dynamic environments with limited support staff
- Requirements
  - Facilities: Not specified
  - Equipment: JSAF can run as a stand-alone system or in a networked environment

- Personnel: System operators familiar with Linux. Would likely require a significant number of support personnel, including meteorological staff to correctly model environmental data feeds that plug into JSAF.
- Required training: Calytrix, a popular provider of consulting and training to the defense training and simulation community, offers two-and-a-half-day and four-day courses to train operators to create scenarios, load entities, and tailor the program for their particular uses, as well as compiling the JSAF platform from source code and enabling various command and control options to tailor the program for end users. Cost is $2,500 per student or $10,000 for on-site training for up to eight individuals.

Notes:
- A technical report sponsored by the AFRL noted the complexity of the software, as well as the large staff and training required to effectively use it. A NATO report also noted the system's poor documentation.
- The NWC finds JSAF a good adjudication tool because it is based on actual data. However, since it is a POR with current data, it can be harder to use for future scenarios. It is an attrition-based model that gives high fidelity for adjudicating force-on-force encounters, but the deterministic nature of the model means there are limitations for use in wargames.
- The NWC also uses JSAF for wargame visualization, but notes that it is very complicated to run.

Joint Seminar Wargaming Adjudication Tool
- Overview
  - Center: DST
  - Tool/approach: jSWAT and jSWAT2
  - Description: Computer-analytical seminar wargame support tool including electronic map and order of battle management with a focus on being able to wargame directly from the planning products and to evaluate different outcomes of events. Planning and wargaming are supported by a variety of adjudication tools including movement, logistics, combat, and ISR operations. Focus is toward the digitization of the manual wargaming process.
  - Government sponsor: DST
  - Developer: DST
- Information
  - Focuses: All
  - Operational level: Variable, depending on focus of the problem. Has been used from tactical to operational levels.
  - Purpose: Analytical seminar wargaming, data collection, planning, and adjudication
  - Forces: All

- Current uses: To understand the causes of failure across capabilities (system, force structures and/or concepts/procedures), contexts, and actions. To identify the feasible scenario space of capabilities. To determine how the force operates on the boundary of success. Determine effectiveness, robustness, and flexibility through explicit consideration of environmental impacts, impact of the enemy, and capability interdependencies.
- Users: Air Power Development Center, Australian Army Experimental Framework, Australian Joint Experimentation Directorate, NPS, and U.S. Defense Systems Analysis Division and HQ1
- Limitations: Simple adjudication models
- Requirements
  - Facilities: Minimal
  - Equipment: Minimum is a laptop computer, but also designed to work with touch screens, if available. Can also work distributed if network is available.
  - Personnel: Standard for analytical seminar wargaming
  - Required training: Light training requirement; includes a training package

Notes:
- NPS has fielded jSWAT in a lab based on a memorandum of understanding. It is also available to the Army, the Dstl, and the Marine Corps.

Joint Wargame Analysis Model
- Overview
  - Centers: CAA, CNA
  - Tool/approach: JWAM
  - Description: JWAM is a manual, computer aided, time-step, human-in-the-loop, force-on-force simulation methodology developed over ten years at the CAA
  - Government sponsor: CAA
  - Developer: CAA
- Information
  - Focuses: Conventional, cyberoperations, and space
  - Operational levels: tactical, operational, and strategic
  - Purpose: Analysis and data collection
  - Forces: All
  - Current uses: Evaluating and comparing COAs to support CCMD operational plans and defense planning scenario development. Testing whether defense planning scenario forces and CONOPS are sufficient. Creating a consensus within the SSA community about what the combat for a scenario would look like; aiding campaign modelers in understanding operational flow and critical campaign events before coding into higher resolutions computer models such as the Joint Integrated Contingency Model. Serving as a test bed for potential operational-level impact of emerging technologies.

- Limitations: Rule set not verified, validated, and accredited; wargame results need to be quantifiable because it is driven by need to go into a model
- Requirements:
  - Facilities: Minimal
  - Equipment: Excel for analysis, Microsoft Access for battle tracker, whiteboards, cork boards
  - Personnel: Operations research analysts familiar with JWAM
  - Required training: No formal training available; requires six months of practice on wargames to become adept at JWAM method

Notes:
- JWAM is purposely manual and paper-based, with pasteboard icons because of the need for it to be reliable when the CAA travels to other locations. The CAA has looked into modernizing it for a number of years, and has considered tools such as light tables, SWIFT, and VAST. For example, SWIFT may be able to automatically record moves and use the JWAM battle tracker. However, CAA believes that automating the JWAM process too much may lead to insufficient consideration of branches and sequels and has thus made a conscious decision to keep it low-tech.
- CAA also uses Microsoft Access for the JWAM "battle tracker" and Excel for game analysis. CAA also uses Google Earth for maps.
- The game process uses two or more opposing teams. It uses free play, and aggregated tactical outcomes for operational-level insights. The model is divided into 13 distinct steps. The emphasis is on Phase II and Phase III of operational campaigns, with 24- or 72-hour time steps.
- The 13 steps are as follows:
  1. Determine Weather
  2. Cyber/Space/EW Operations
  3. ISR Operations
  4. Integrated Air Defense System Allocation
  5. Strategic Deep Strike Missions
  6. Determine Air Superiority
  7. Strategic Deployment
  8. Logistical Sufficiency Check
  9. Naval Combat
  10. Tactical Deep Strike Missions
  11. Ground Combat
  12. Post Combat
  13. Post-turn Hot Wash.[19]

---

[19] Mahoney, 2016a, p. 7.

Large-Format Printers
- Overview
  - Centers: CNA, CSL, NWC, RAND
  - Tool/approach: Large-format printers
  - Description: Large- or wide-format printers are generally accepted as any printers that allow a maximum width of 18–60 inches
  - Government sponsor: None
  - Developer: Various manufacturers, including Canon, HP, and Xerox
- Information
  - Focuses: Presentation and visualization
  - Operational levels: All
  - Purpose: Large-format printers are useful for printing maps, posters, and various visualization aids for use during games
  - Forces: Any
  - Current uses: High-quality prints of maps and game pieces are useful for increasing game immersion
  - Limitations: High potential cost of $5,000–$50,000
- Requirements
  - Facilities: Room for printer
  - Equipment: Specialized large-format printer, paper, and ink
  - Personnel: Minimal personnel requirements
  - Required training: Minimal

Map Aware Non-Uniform Automata
- Overview
  - Center: NPS
  - Tool/approach: MANA
  - Description: MANA is an agent-based model that allows the exploration of a wide variety of issues with minimal setup time. Its main advantages are simplicity and ease of use. Many naturally occurring phenomena are too complex to be captured accurately in models, and highly detailed models of these phenomena are necessarily arbitrary. Therefore, the NPS advocates using simple models, such as MANA.
  - Government sponsor: New Zealand Defence Technology Agency
  - Developer: New Zealand Defence Technology Agency
- Information
  - Focus: Simulation
  - Operational level: Unit- and tactical-level simulations where individual behavior can significantly affect outcomes. Simulating security, stability, or reconstruction operations where forces may encounter civil discontent and violence are potential uses of this tool.

- Purpose: This agent-based M&S tool can be useful for exploratory analysis during the early stages of advanced concept development. It can point to issues and generate questions that can then be explored with higher fidelity models.
      - Forces: Ground (current uses)
      - Current uses: Unclear how actively this tool is used. It was previously used by the Marine Corps Warfighting Lab and foreign militaries.
      - Limitations: MANA is meant to provide general behavioral insights. Quantitative results may not be useful. Limitations in sensing, communication, elevation, and weapon models can make the tool inadequate for use in certain combat simulations. Some sophisticated behaviors may be unattainable with MANA. Some of these limitations have been addressed for future versions of the software.
- Requirements
      - Facilities: Minimal
      - Equipment: Unknown
      - Personnel: The model's simplicity can allow new users to set up simple simulations in a few hours
      - Required training: Not specified
      - Cost: Unknown

Notes:
- Despite MANA's relative ease of use, instructors at the NPS emphasized that there is no simple plug-and-play combat simulator for educational purposes.

Map Exercise
- Overview
      - Center: TRAC
      - Tool/approach: MAPEX
      - Description: The planning of forces or a scheme of maneuver, development of a synch matrix, and/or COA development as part of the military decisionmaking process. MAPEX wargames are useful for exploring the tactical impact of a specific vignette. TRAC uses MAPEXes extensively during scenario development.
      - Government sponsor: TRAC
      - Developer: TRAC
- Information
      - Focus: Conventional
      - Operational level: Tactical and operational
      - Purpose: Visualization, data development, scenario development, and plans development
      - Forces: Ground, air, and cyberoperations
      - Current uses: Used to develop Army scenarios that underpin Analysis of Alternatives
      - Limitations: Primarily used for scenario development

- Requirements
  - Facilities: Room or auditorium with map on floor
  - Equipment: Large floor map and manual counters for forces
  - Personnel: Fifteen scenario developers, 30 SMEs
  - Required training: Unclear

Notes:
- Stages to running a typical TRAC MAPEX: planning and preparation (four weeks), execution (one week), and postevent processing (two to three weeks)
- Typically involves 15 scenario developers in planning, preparation, and postevent processing phases. Also, about 30 Centers of Excellence participants to act as SMEs.

Massive Multiplayer Online Wargame Leveraging the Internet
- Overview
  - Center: NPS Modeling, Virtual Environments, and Simulation Institute
  - Tool/approach: Massive Multiplayer Online Wargame Leveraging the Internet
  - Description: Online brainstorming platform for running discussions with strict format. Players are able to play 140-character "idea cards," which other players then respond to. Platform includes a blog and other means of getting game updates to participants.
  - Government sponsor: NPS
  - Developer: NPS
- Information
  - Focus: Any
  - Operational levels: All
  - Purpose: Design and data collection
  - Forces: Any
  - Current uses: Navy wargaming support
  - Limitations: Card structure best suited to brainstorming; platform lacks flexibility for other types of game interaction
- Requirements
  - Facilities: Minimal
  - Equipment: Internet
  - Personnel: Support from NPS Modeling, Virtual Environments, and Simulation Institute to use game platform
  - Required training: System requires little training to use as a player

Notes:
- More information is available at the Massive Multiplayer Online Wargame Leveraging the Internet portal.

Matrix Games
- Overview
  - Centers: CORA, CNA, CSL, Dstl, NUWC, RAND, UK Defence Academy
  - Tool/approach: Matrix games
  - Description: Means of structuring gameplay between multiple teams that enables players to weigh in on the likely outcome of other teams' actions. Players attempt an action and explain why they think their action will be successful; others weigh in on why they think the action will not be successful.
  - Government sponsor: N/A
  - Developer: Chris Engle, popularized by John Curry
- Information
  - Focus: Any
  - Operational level: Any
  - Purpose: Game design and adjudication
  - Forces: Any
  - Current uses: Substitute for seminar-style games, particularly when gaming emerging topics with limited expertise to inform adjudication. Topics that matrix games have been used for include cyberoperations, ISIL, Russian aggression in the Baltics, DoD's Third Offset Strategy, and urban warfare.
  - Limitations: Method requires a confident facilitator to stay on track, and even very productive matrix games can appear unstructured during the event. However, it will not provide detailed quantitative data.
- Requirements
  - Facilities: Equivalent to a seminar-style game or tabletop exercise
  - Equipment: Minimal
  - Personnel: Requires trained facilitators to execute, as well as scenario development
  - Required training: Helpful to both read a description of the approach and to observe a game in this style prior to running

Notes:
- A key text by John Curry and Tim Price includes both a description of the Matrix game approach and several sample games[20]
- The basic concept of matrix games is flexible, and has been adapted to meet different requirements. For example, NUWC's narrative games are a variation of this approach, and RAND has run hybrid RFLEX-matrix games with some success. Several systems of matrix gaming exist.
- Adjudication may also use a combination of expert judgment, umpiring, and weighted probability

---

[20] John Curry and Tim Price, *Matrix Games for Modern Wargaming Developments in Professional and Educational Wargames: Innovations in Wargaming*, Vol. 2, Morrisville, N.C.: Lulu Press, 2014.

- One wargaming session at the October 2016 MORS wargaming special meeting focused on matrix gaming and was led by a participant from the UK Defence Academy.

MaGCK Matrix Game Construction Kit

- Overview[21]
  - Center: Dstl
  - Tool/approach: MaGCK Matrix Game Construction Kit
  - Description: Kit contains matrix game rules, example matrix game and scenarios, map tiles, blank tokens, discs, tracking mats, dice, and printable templates
  - Government sponsor: UK MOD
  - Developer: The Game Crafter
- Information
  - Focus: Varies
  - Operational level: Varies, but provided map tiles support the tactical level
  - Purpose: To provide materials to support matrix games
  - Forces: Varies, but has an emphasis on military and paramilitary symbols
  - Current uses: Supports the creation of a variety of matrix games
  - Limitations: Limited size and finite materials. Included map tiles are at the tactical level.
- Requirements
  - Facilities: Minimal
  - Equipment: Purchase of game kit
  - Personnel: Personnel familiar with matrix gaming
  - Required training: Training on matrix gaming

Modified Commercial Board Games

- Overview
  - Centers: CAPE, Central Intelligence Agency, CGSC, CNA, CSL, NPS, ONI, TRAC
  - Tool/approach: Modified versions of commercial board games, including After the Holocaust; Algeria: the War for Independence; Axis and Allies; Drive on Paris; En Guard; Friedrich; Guerilla Checkers; Hanabi; Harpoon; Kriegsspiel (1821 rules); and Ultimate Werewolf
  - Description: Commercially available games for educational or analytical use. For all but the simplest games, modification is generally needed to make rule sets more accessible and create a playable game to address the educational requirement.
  - Government sponsors: Various

---

21 See The Game Crafter, "MaGCK Matrix Game Construction Kit," webpage, undated.

- Developer: Commercial entities including Avalon Hill, Fiery Dragon Productions, and GMT Games
- Information
  - Focus: Varies
  - Operational level: Varies
  - Purpose: Education and experiential learning on topics, including COA development, principles of COIN, and ethics and decisionmaking
  - Forces: Varies
  - Current uses: Supports education and training at the CGSC and ONI, concept development for CAPE. The CSL is currently collaborating with the developer of Breaking Chains to simplify its gameplay and convert the game into a matrix-style wargame.
  - Limitations: Games are designed for entertainment, and may have important simplifications and inaccuracies. Many games are too complicated to be well suited to use in classroom.
- Requirements
  - Facilities: Minimal
  - Equipment: Purchase of commercial games
  - Personnel: Familiarity with commercial game rules
  - Required training: Familiarity with the game prior to use can require substantial investment in learning games

Narrative Gaming
- Overview
  - Center: NUWC
  - Tool/approach: Narrative gaming
  - Description: Method for running seminar-style games aimed at developing innovative concepts. The goal of this method is to explore the problem by designing a game that represents the context and resources that shape realistic solutions. Players receive an initial scenario and are asked to present a preferred COA in a written paragraph that states what the players want to do and the rationale for why they believe it is plausible. These rationales emerge organically in small group discussion, but forcing the group to provide written input provides for a more structured response and decreases data collection burden on game administers.
  - Government sponsor: NUWC
  - Developer: NUWC
- Information
  - Focus: Any
  - Operational levels: All
  - Purpose: Design, adjudication, and data collection

- Forces: Any
- Current uses: Navy wargaming support
- Limitations: Methodology requires strong facilitation in order to provide structure, otherwise it is subject to many of the same limitations as classical seminar games
- Requirements
  - Facilities: Space for cell-level conversations, ideally with whiteboards or other means of capturing diverse types of information in a flexible manner
  - Equipment: None
  - Personnel: Game designers (experience with seminar and matrix gaming a plus), experienced facilitators, trained note takers
  - Required training: Facilitation training and observation of games of this type would be helpful but are likely not required for success

Notes:
- NUWC has found it advantageous to bring participants into the scenario development process in the lead-up to the exercise in order to generate buy-in from key stakeholders
- Note takers should pay particular attention to the ideas that are generated in the game but are not selected for the group, as well as the arguments for and against them
- NUWC used Virtual Worlds as a visualization tool to display the results of its game.

National Security Policy Analysis Forum Seminar Game
- Overview
  - Center: CASL
  - Tool/approach: National Security Policy Analysis Forum Seminar Game
  - Description: Single cell, three-move seminar game. In each move, a scenario is presented, with a list of discussion question for participants. There is no adjudication between moves. The format is well suited for interagency games or other areas were information sharing and relationship building is important to future problem solving.
  - Government sponsor: OSD Policy
  - Developer: CASL
- Information
  - Focus: Any
  - Operational level: Strategic
  - Purpose: Design
  - Forces: Any
  - Current uses: Approximately five games per year for OSD Policy offices
  - Limitations: Fixed scenario is not responsive to player actions. Additionally, the structure of the discussion is highly dependent on facilitators' ability to

focus players toward reaching decision points rather than just discussing the problem.
- Requirements
  - Facilities: Single room with conference table
  - Equipment: Computer for slides
  - Personnel: Three to four designers to produce scenario and manage game execution
  - Required training: Knowledge of international relations and security topics and observation of the game design process

Naval Postgraduate School Analytic Wargaming Course
- Overview
  - Center: NPS
  - Tool/approach: NPS analytic wargaming course
  - Description: Eleven-week course offered by the NPS. The basic skills required to design, develop, conduct, and analyze a wargame are taught in the first six weeks of the course through a series of interactive lectures and practical exercises. Weekly labs are focused on putting into action the skills taught in the weekly lectures. The capstone wargaming project allows the students to construct an actual wargame through leveraging those skills accumulated in the first six weeks of the course. A DoD sponsor with a real-world problem will provide the topic of the wargame that serves as the end-of-course project. A one-week Mobile Execution Team version can also be offered on the road.
  - Government sponsor: NPS
  - Developer: NPS
- Information
  - Focuses: Basics of design, development, execution, and analysis of wargames
  - Operational levels: All
  - Purpose: To educate students to know and differentiate between the classes and types of wargames, understand how wargaming fits into campaign analysis, and understand data collection and management plans and their role in wargame design. The course also helps students understand nonquantitative metrics often used in wargames: observations, insights, findings and how analysis (linked with the sponsor's objective and issues) drives the wargame's design. This course is designed to bring students from a basic understanding of what a wargame is to learning the skills of an apprentice wargame designer and analyst.
  - Forces: All
  - Current uses: Applied education—students will design, develop, conduct, and analyze a wargame for a DoD or defense partner sponsor
  - Limitations: Limited complexity and depth to fit course length requirements and student experience

- Requirements
  - Facilities: Briefing rooms, separate spaces for blue, red, and white cells
  - Equipment: Game tools can be designed with standard MS Office. May also need plotting equipment, map printing capabilities, and three-dimensional printers for game pieces.
  - Personnel: One instructor; course also benefits from real-world clients that sponsor student-designed games
  - Required training: One week to 11 weeks
  - Cost: $30,000–$60,000 for one-week course, depending on level of customization and travel costs

Naval War College Web Applications
- Overview
  - Center: NWC Wargaming Department
  - Tool/approach: NWC Web applications
  - Description: Custom build GUI and modules to support game execution and data collection tasks (a more extensive list of capabilities in supplemental briefing)
  - Government sponsor: NWC
  - Developer: NWC
- Information
  - Focus: Any
  - Operational levels: All
  - Purpose: Analysis, data collection, and visualization
  - Forces: Naval, ground, and air
  - Current uses: Navy wargaming support
  - Limitations: Modules need to be configured for each game by a technical expert
- Requirements
  - Facilities: Requires appropriate network certification to develop and deploy in-house software
  - Equipment: Computer network (either wired or virtual), including front-end terminals and servers. Network specification and server requirements determined by number of simultaneous players. The NWC is moving to virtual machines using a microcloud and multiblade servers, with on-demand workstations.
  - Personnel: C Sharp and Java script developers able to interface with game designers to properly integrate and expand tools. NWC currently has a staff of five responsible for development and managing execution. Staffing requirements are not scalable (i.e., one fewer developer does not mean one-fifth the games as a range of skills are required).

– Required training: Tools are designed to require minimal player training (at most a half day) to use; however, deploying a configuration requires full-time software developers, who could be introduced to the existing modules fairly easily, but there is not an established process for training

Notes:
- Need a network that is scalable and movable to allow for frequent reconfigurations based on game needs
- Requires long lead time: three months of development, three months testing, and three months for final tuning of the configuration. Time can be shortened in some cases.
- Almost one-third of the labor goes to information assurance
- A hard cutoff date for development concepts (the "Good Idea Fairy") must be established and enforced. After that date, bug fixes or feature fixes should be the only changes allowed; otherwise, game designers/developers will have web development "requirements" constantly added, which vastly increases the risk of something going awry.

Operations Research Textbooks: Methods of Operations Research and Search and Screening

- Overview
  - Center: CNA
  - Tool/approach: Operations research textbooks: *Methods of Operations Research* and *Search and Screening*
  - Description: General-purpose textbooks that provide fundamental OR methods with various military applications
  - Government sponsor: Operations Evaluations Group, Office of the Chief of Naval Operations
  - Developers: Philip M. Morse and George E. Kimball (*Methods of Operations Research*); Bernard Osgood Koopman (*Search and Screening*)[22]
- Information
  - Focuses: Adjudication and analysis
  - Operational level: Tactical
  - Purpose: Wargames, especially those involving naval forces, may entail analysis and calculation of detection probabilities, chances of success for various search patterns and scenarios, and gunnery and bombardment problems for which these texts provide useful explanations and methods for measuring the effectiveness of certain tactics
  - Forces: All
  - Current uses: Tools for analysis of game results

---

[22] Morse and Kimball, 2003; Koopman, 1999.

- – Limitations: Books are old, methods may be outdated, and one appears to be out of print, though both can be found online
- Requirements
  - – Facilities: None
  - – Equipment: Computer for calculations
  - – Personnel: Individual to run analysis and adjudicate results
  - – Required training: Familiarity with OR and necessary mathematics background

Notes:
- Other OR textbooks would be equally useful, but these texts were explicitly identified during interviews.

Commercial Personal Computer Games

- Overview
  - – Center: LeMay Center for Doctrine Development and Education
  - – Tool/approach: Commercial PC games
  - – Description: PC-based computer games provide a wide range of strategic and tactical games, spanning various formats and settings in terms of time period and depth of detail. These games can be viewed as modern evolutions of strategic board games like Axis and Allies. The key difference between board games and PC-based games is the live adjudication by the software. Examples of PC games include *Empire*, *Rise of Nations*, *Rome: Total War*, *StarCraft*, and *Warcraft*.
  - – Government sponsor: N/A
  - – Developer: Various commercial vendors
- Information
  - – Focus: Education
  - – Operational levels: All
  - – Purpose: Commercial PC games are used to familiarize students with different types of gaming, ranging from the strategic to the tactical
  - – Forces: All
  - – Current uses: Commercial PC games allow participants to engage in real-time strategy exercises in various formats and contexts. They are often used to expose students or first-time gamers to wargaming concepts and gameplay.
  - – Limitations: PC-based games are proprietary software, which means these games cannot be modified or changed and require a fee to purchase
- Requirements
  - – Facilities: N/A
  - – Equipment: PC
  - – Personnel: N/A
  - – Required training: Commercial games are designed to be accessible to the general public. Players often learn as they play.

Peace Support Operations Model
- Overview[23]
  - Center: Dstl
  - Tool/approach: PSOM
  - Description: PSOM is a computer-assisted wargame designed to represent the full range of civilian and military activity in a stabilization scenario. The wargame provides a method to analyze the stabilization problem space within the context of a wider cross-government response.
  - Government sponsor: Dstl
  - Developer: Dstl
- Information
  - Focus: Civilian population in stability operations
  - Operational level: Campaign
  - Purpose: Adjudication
  - Forces: Stabilization, COIN, and irregular warfare
  - Current uses: Decision support for examining questions relating to stabilization, COIN, and irregular warfare
- Requirements
  - Facilities: Minimal
  - Equipment: Computer
  - Personnel: Personnel training in PSOM
  - Required training: Information not readily available

Professional Facilitation Training
- Overview
  - Center: CASL
  - Tool/approach: Professional facilitation training
  - Description: Facilitation training designed for consultants working with clients to move though a strategic planning process. This approach gives formal facilitation training to wargame designers and facilitators. Training covers techniques to structure the conversation both generally (e.g., how to get people back on track when they are going down a rabbit hole) and specific (e.g., exercises for brainstorming). Some of the later approaches are similar to those in the *Applied Critical Thinking Handbook*. On the high end of facilitation training, there may be multiple recorded observations of trainees, formal reports, and a professional certification.
  - Government sponsor: CASL
  - Developer: Leadership Strategies

---

[23] See Howard Body and Colin Maston, "The Peace Suppose Operations Model: Origins, Development, Philosophy and Use," *Journal of Defense Modeling and Simulation*, Vol. 8, 2010, pp. 69–77.

- Information
  - Focuses: All
  - Operational levels: All
  - Purpose: Adjudication
  - Forces: All
  - Current uses: CASL; some other centers have one or two individuals who have received similar training (e.g., J-8 SAGD and RAND)
  - Limitations: While many facilitation approaches are general enough to apply to a wide range of gaming approaches, commercially available trainings are not designed with facilitating games in mind. As a result, the trainee must work to determine which approaches are appropriate for any given wargame.
- Requirements
  - Facilities: Minimal. Training can either be done on-site with a visiting trainer or off-site at companies that specialize in facilitation training.
  - Equipment: Butcher block paper or whiteboard, curriculum materials and handouts
  - Personnel: Professional facilitators with experience in leading facilitation training at the low end; professional society accreditation on the high end
  - Required training: Focus of training is presenting different approaches and providing a time to practice. Course formats vary widely.
  - Limitations: Expense. Also, the facilitation skills required for different types of games differ, but available facilitation training is general and not tailored to wargaming.

Notes:
- Leadership Strategies was the company used by CASL to provide facilitation training for its wargaming staff[24]
- Local professional societies, such as the Mid-Atlantic Facilitator Network, offer training classes on facilitation topics and access to a network of facilitators[25]
- Other professional societies, such as the International Association of Facilitators, offer certification in facilitation.[26]

R
- Overview[27]
  - Center: CAPE
  - Tool/approach: R

---

[24] See Leadership Strategies, homepage, undated.

[25] See, e.g., Mid-Atlantic Facilitator Network, homepage, undated.

[26] See, e.g., International Association of Facilitators, homepage, undated.

[27] See R Foundation, "The R Project for Statistical Computing," webpage, undated.

  - – Description: A programming language used for statistical data analysis and visualization
  - – Government sponsor: N/A—open-source software
  - – Developer: R Development Core Team
- Information
  - – Focus: Any
  - – Operational levels: All
  - – Purpose: Analysis and visualization
  - – Forces: Naval, ground, and air (current uses)
  - – Current uses: Wargame analysis
  - – Limitations: R is an open-source, community-developed programming language, so commercial support may be limited
- Requirements
  - – Facilities: None
  - – Equipment: R requires a PC running Linux, Mac OSX, or Windows
  - – Personnel: A single individual can use R to perform statistical analysis of data and produce visualizations
  - – Required training: R is a relatively complex programming language with a very large number of packages and built-in capabilities. There may be a significant learning curve for personnel more familiar with other statistical software packages such as SAS, the SPSS, and Stata.

Notes:
- CAPE uses R for data analysis in wargames
- OAD uses R as an analyst tool
- R has an Army Network Command Certificate of Networthiness and is approved for use on Army networks.

RAND Framework for Live Exercises
- Overview
  - – Center: RAND
  - – Tool/approach: RFLEX
  - – Description: A "live wargame" methodology consisting of a blue team and a red team usually (but not always) working around a common map with a typically transparent adjudication process. Adjudication is through counters and die rolls.
  - – Government sponsors: Various
  - – Developer: RAND
- Information
  - – Focus: Exploratory analysis
  - – Operational level: Operational, multibattalion combat and higher

- – Purpose: Used to reveal decision points, broad strategies, and how participants think about and deal with unexpected outcomes. The transparent adjudication process encourages stakeholder buy-in.
  - – Forces: Army, Navy, Air Force, and Marine Corps
  - – Current uses: Most recently used by multiple services (including the Air Force and Marine Corps) to game various U.S.-Russia scenarios in the Baltics
  - – Limitations: Does not adapt well to tactical-level gaming, where the simplistic rule set and combat simulation limit usability
- • Requirements
  - – Facilities: Conference table for map, room for 9 to 12 players (nominal) and adjudicators
  - – Equipment: Map printers for the game board, presentation equipment to allow team members to collaborate and develop CONOPS
  - – Personnel: Depends on size and dimensionality of game. In games involving multiple force types (ground, air, sea, cyberoperations, space), expertise will be needed in each area for adjudication. Note takers are also required to track the adjudication results.
  - – Required training: The simple rule set allows for relatively quick player training. However, expertise is needed for the design, execution, and adjudication of these games, but the rules and the game are meant to be easily adaptable to a wide range of scenarios.

Rapid Campaign Analysis Toolset

- • Overview[28]
  - – Centers: Dstl and Cranfield University
  - – Tool/approach: RCAT
  - – Description: Manual simulation framework for designing a wide variety of workshops and wargames. The toolset consists of a set of baseline mechanisms and rules for issues like movement combat, logistics for use with a map and counters, and more developed rule sets covering particular topics of interest. However, the system is often combined with expert judgment, and allows for considerable flexibility based on the nature of the conflict.
  - – Government sponsor: Dstl
  - – Developer: LBS Consultancy and Cranfield University
- • Information
  - – Focuses: All
  - – Operational level: Generally operational, but the approach could likely be adapted to other levels of war

---

28 See Rex Brynen, "Playtesting RCAT," PAXsims (blog), November 30, 2015.

- Purpose: Game design, COP, and adjudication
- Forces: All
- Current uses: Used by Dstl and Cranfield University for wargame design
- Limitations: Framework is highly flexible, which means it can be applied to problems that may not be tractable to gaming or may be used by a designer who is not able to calibrate the design to meet game objectives
- Requirements
  - Facilities: Minimal
  - Equipment: Map and counters
  - Personnel: Relatively low effort, but requires at least a designer to adapt the rules to the event purpose and a facilitator to guide discussion
  - Required training: Minimal, but facilitation training/experience is likely necessary for success

SharePoint
- Overview
  - Center: NWC
  - Tool/approach: SharePoint
  - Description: Organizations use SharePoint to create websites that act as a secure place to store, organize, share, and access information
  - Government sponsor: N/A; COTS product
  - Developer: Microsoft
- Information
  - Focus: Knowledge management
  - Operational levels: All
  - Purpose: Can be used by players during gameplay for knowledge management, information sharing, and communication
  - Forces: Any
  - Current uses: Information sharing and knowledge management during wargames
  - Limitations: Potential permissions and licensing issues for multiple networks. SharePoint also has limited flexibility in its interface.
- Requirements
  - Facilities: Workspace for computers
  - Equipment: For use, computer with a web browser; for installation, Windows Server 2012 with 16 gigabytes of random-access memory, 64-bit multicore processor, 80 gigabytes of hard drive space
  - Personnel: Minimal
  - Required training: A number of online/virtual solutions are offered through Microsoft. However, one of the benefits of using SharePoint is that players have likely used SharePoint in other contexts and are familiar with the software's capabilities.

— Cost: Can be purchased through a cloud-based service for $5–$10 per person or as an enterprise server solution where the software is installed on a local server and managed internally

Simulation-Based Analyst Training

- Overview
  - Center: ONI
  - Tool/approach: SimBAT
  - Description: A program within ONI that serves both training and analysis objectives
  - Government sponsor: ONI
  - Developer: ONI, using commercial board games
- Information
  - Focuses: Training and analysis
  - Operational level: Operational and strategic
  - Purpose: To provide a multidimensional, "all-sensory" training model to junior analysts that familiarizes them with the staff planning process and strategic decisionmaking. Using slightly tailored versions of classic commercial board games, such as Axis and Allies, students are presented with cost-benefit trade-offs that they must weigh as they consider the roles and missions of various forces, how force structure aligns to particular strategies, and how national objectives align with enemy threat postures and capabilities. SimBAT emphasizes situation assessment and the structured analysis that they will utilize in their jobs.
  - Forces: Naval, ground, and air (current uses)
  - Current uses: Junior analyst training
  - Limitations: Requires skilled facilitation and an ability to adapt and streamline commercial board games to accomplish desired learning objectives
- Requirements
  - Facilities: Separate rooms for opposing teams to work in during the planning process, and a room for both teams to participate across a common game board
  - Equipment: Minimal. Board game campaign models can be made with very limited resources and are of low cost. Because there is no software, gameplay is flexible and rules can be altered quickly.
  - Personnel: Experienced facilitator familiar with various elicitation techniques to improve student engagement and learning. One individual can run the simplest games, where the goal is to emphasize the uncertainty in time aspects of decisionmaking and the focus is on the planning process and force composition. For more dynamic campaigns where uncertainty across space is emphasized, up to five individuals may be required as adjudicators and embeds in red and blue cells.
  - Required training: Less than a day for participants to understand rules and objectives

Soft Systems Methodology/Strategic Design Methodology
- Overview
  - Center: SOCOM Wargame Center
  - Tool/approach: SSM/strategic design methodology
  - Description: This tool is based on our use of design thinking as an application of Qualitative Research that uses this methodology to achieve the tenements of design thinking. A method of inquiry, learning, and analysis, often used in business, when there are competing views on the definition of the problem. Unlike systems engineering or OR, which can be used to find an effective means to achieving a well-defined and agreed-upon objective, SSM is useful when the objective itself may be contentious. SSM assumes that in any organization, stakeholders will have divergent interpretations of the same events driven by varying goals and experiences.
  - Government sponsor: N/A
  - Developer: Ascendance International/Sam Bass, Mike Wertz, and Reb Yancey, based upon hybrid of SSM approaches such as those authored by Peter Checkland and Peter Senge (among others) and facilitated in a Socratic methodology
- Information
  - Focuses: Facilitation and analysis
  - Operational levels: All
  - Purpose: SSM/strategic design methodology provides a structured way to bring SMEs together to think systematically about a range of problems by creating a venue where a range of interpretations can be identified, their assumptions made explicit, and various trade-offs examined. This process of inquiry and discourse can be captured and packaged into presentations for senior leaders.
  - Forces: SOF, joint, interagency, intergovernmental, and multinational
  - Current uses: Utilized in seminar-style wargames centered around various issues
  - Limitations: Not an adjudicated game
- Requirements
  - Facilities: Conference room in which SMEs can present and discuss
  - Equipment: No hard requirements; effective use of SSM is facilitated by whiteboards, tables, and visual presentation aids
  - Personnel: Mission area SMEs for facilitation though a professional full-time facilitator may be worthwhile. OR analysts trained in "soft" OR methods such as SSM.
  - Required training: Facilitation and subject matter expertise. No training required for participants.

Notes:
- SSM was first developed in the 1970s by Peter Checkland
- SSM is sometimes categorized as a "soft" method in U.S. OR, a problem-structuring method in UK OR, and a judgment-based method in NATO OR.

Spreadsheet Tools

- Overview
  - Centers: AFRL, LeMay Center for Doctrine Development and Education
  - Tool/approach: Spreadsheet tools
  - Description: Digital documents in which data is organized via rows and columns, which allows for various configurations of data and calculations. Typical spreadsheet tools include Excel and Google Spreadsheets, the latter a web-based program.
  - Government sponsor: N/A
  - Developers: Microsoft, Google, and others
- Information
  - Focuses: Design and data collection
  - Operational level: N/A
  - Purpose: To provide a means to organize data in a meaningful manner and provide calculation tools
  - Forces: N/A
  - Current uses: Provide convenient media to record and manipulate various data sets, whether for wargame design, data collection, or analysis
  - Limitations: Obtaining and maintaining stand-alone system imposes considerable bureaucratic and fiscal expense, as the center must assume the risk for the system
- Requirements
  - Facilities: N/A
  - Equipment: Computer and access to software
  - Personnel: N/A
  - Required training: Minimum training for daily use. For more in-depth training, there are a plethora of online tutorials for spreadsheet tools.

Stand-Alone Computer Network

- Overview
  - Centers: NUWC, NWC
  - Tool/approach: Stand-alone computer network
  - Description: Stand-alone networks for testing software and running games that are separate from the usual government network approval processes. These are critical for allowing more experimental software and approaches to be tested before going through the whitelisting process. Centers currently use these networks at both unclassified and classified levels.
  - Government sponsor: NUWC, NWC
  - Developer: NUWC, NWC
- Information
  - Focuses: All
  - Operational levels: All

- Purpose: All
- Forces: All
- Current uses: Provides a space for new software to be used in game design, execution, and analysis without posing a risk to the main system. Allows for gaming centers to try out potential software with trial versions before actual purchase or installation.
- Limitations: Obtaining and maintaining stand-alone system imposes considerable bureaucratic and fiscal expense, as the individual center must assume the risk for the system

- Requirements
  - Facilities: Varies
  - Equipment: Varies
  - Personnel: Required dedicated technical staff to stand up and maintain the network, including both routine and game-specific configuration
  - Required training: Varies, but personnel must meet DoD 8570 certification requirements; networks must adhere to Defense Information Systems Agency accreditation requirements

Standard Wargaming Integration and Facilitation Tools
- Overview
  - Centers: CAPE, CNA
  - Tool/approach: SWIFT
  - Description: A software gaming engine for designing, playing, adjudicating, and analyzing turn-based games. SWIFT is an environment in which to design, play, visualize, record, play back, and analyze wargames. It is also designed to permit adjudication via N number of internal adjudicators (to include a manual), as well as interfaces to external adjudicators (e.g., JWAMs Battle Tracker).
  - Government sponsor: CAPE, M&SCO, OSD Operational Energy
  - Developer: CAPE
- Information
  - Focuses: Knowledge management, visualization, recordation, adjudication, automation, and data analysis
  - Operational levels: All (agnostic)—existing wargames within SWIFT from the strategic to tactical levels
  - Purpose: SWIFT takes the computer-based wargaming environment with requisite design accessibility, limiting the need for software developers and placing significant automation power in the hands of the analyst, event facilitator, and wargame designer
  - Forces: All

- Current uses: Wargame visualization, recordation, and adjudication for current year and out-year scenarios. Use by CCMDs in bilateral and multilateral wargames; recent emphasis on logistics wargaming.
  - Limitations: Software interface supports any wargame, but is not designed to a specific wargame or purpose. This has been mitigated somewhat through extensive use and enhancements in the last few years.
- Requirements
  - Facilities: Minimal facility requirements. Room for computer workstations and projection/large-screen displays.
  - Equipment: As a Java application, SWIFT can run on most current laptops and PCs
  - Personnel: Personnel requirements range from a trained analyst to a software developer, depending upon the complexity of the automation desired
  - Required training: Learning curve varies based on the role, but a point-and-click interface is available for players to quickly get up to speed

Statistical Package for Social Sciences

- Overview
  - Center: NWC
  - Tool/approach: SPSS
  - Description: A predictive analytics software tool that allows analysis, data management (case selection, file reshaping, creating derived data) and data documentation. SPSS's wide array of statistical analysis and modeling tools are accessible through both a GUI and a proprietary programming language.
  - Government sponsor: N/A; COTS product
  - Developer: IBM
- Information
  - Focuses: Knowledge management and analysis
  - Operational levels: All
  - Purpose: Thematic/sentiment analysis, quantitative analysis, predictive modeling, factor/cluster analysis
  - Forces: Naval, ground, and air (current uses)
  - Current uses: Navy wargaming data analysis
  - Limitations: High individual license costs. Potential permissions and licensing issues for multiple networks.
- Requirements
  - Facilities: Minimal facility requirements; workspace for computers only
  - Equipment: Client/server architecture. Wide range of Linux, Mac, and Windows platforms supported.
  - Personnel: Individuals experienced in both social science research methods and the SPSS software

- Required training: A wide range of fee-based instructor-led online and in-person training is available from IBM and third-party vendors covering all aspects of SPSS. Videos and other online resources are available free of charge. SPSS is a sophisticated analysis tool that requires knowledge of both the software and the underlying theory to use effectively.
- Cost: $1,170–$11,300 per user per year, depending on software suite

Notes:
- OAD uses SPSS as an analyst tool.

Structured Analytic Techniques
- Overview
  - Centers: Intelligence community, RAND
  - Tool/approach: *A Tradecraft Primer: Structured Analytic Techniques for Improving Intelligence Analysis* and *Structured Analytic Techniques for Intelligence Analysis*, 2nd ed.
  - Description: Handbooks of structured analytic techniques that the intelligence community uses to improve intelligence analysis. The three types of classes of techniques discussed are diagnostic techniques, contrarian techniques, and imaginative thinking techniques.
  - Government sponsor: U.S. intelligence community
  - Developer: U.S. intelligence community; commercial version also available
- Information
  - Focuses: All
  - Operational levels: All
  - Purpose: Visualization
  - Forces: All
  - Current uses: Used by the U.S. intelligence community. RAND has also used SATs to support workshops and wargames.
  - Limitations: Designed for the intelligence community
- Requirements
  - Facilities: Minimal
  - Equipment: Minimal
  - Personnel: People trained in SATs
  - Required training: Some organizations certify SAT facilitators

Notes:
- Government version: U.S. Central Intelligence Agency, *A Tradecraft Primer: Structured Analytic Techniques for Improving Intelligence Analysis*[29]

---

[29] U.S. Central Intelligence Agency, 2009.

- Commercial version: Richard J. Heuer, Jr., and Randolph H. Pherson, *Structured Analytic Techniques for Intelligence Analysis*, 2nd ed.[30]
- SATs, red teaming, "soft operational analysis" (or judgment-based operational analysis), and problem-structuring methods from OR share common methods
- Some consider wargaming an SAT in the category of imaginative techniques.

Synthetic Staff Ride: Mindanao
- Overview
  - Center: TRAC
  - Tool/approach: SSR: Mindanao
  - Description: Low-cost gaming tool through which players are presented with a complex environment, expected to create a strategic plan, and challenged to complete objectives through negotiation and risk mitigation. It is a cooperative, multisided leadership board game that covers shape and deterrence operations in the present-day Philippines. It deals with Phase 0 and the effect of diplomatic, information, military, and economic activities on PMESII-PT considerations. The focus is on developing leadership competencies as outlined in Army doctrine.
  - Government sponsor: TRAC
  - Developers: CNA, TRAC
- Information
  - Focuses: Counterterrorism, economic, interagency, and stabilization
  - Operational level: Tactical, operational, and strategic
  - Purpose: Training game
  - Forces: Ground
  - Current uses: Audiences include the Army's Captains Career Courses, the CGSC, Pre-Command Courses, and MCU
  - Limitations: Leadership development rather than training
- Requirements
  - Facilities: No special facilities required
  - Equipment: All necessary equipment in board game box
  - Personnel: One moderator familiar with the game
  - Required training: Familiarity with both the game and Army leadership doctrine

Notes:
- Six-turn game that requires four hours to play
- Up to seven factions may be played: Abu Sayyaf, the Moro National Liberation Front, the New People's Army, the government of the Republic of the

---

30 Richard J. Heuer, Jr., and Randolph H. Pherson, *Structured Analytic Techniques for Intelligence Analysis*, 2nd ed., Thousand Oaks, Calif.: CQ Press, 2015.

Philippines, the U.S. military, the U.S. Department of State, and nongovern-mental organizations
- The included observer's guide allows an observer to score participant leadership behavior based on the Army Officer Evaluation Reporting System
- TRAC has 50 copies of the game left
- ARCIC uses a game inspired by SSR: Mindanao to familiarize newcomers to staff organizations and activities.

Systemic Operational Design
- Overview[31]
  - Centers: Operational Theory Research Institute, SOCOM Wargame Center, TRADOC
  - Tool/approach: Systemic Operational Design (SOD)
  - Description: A systems theory approach to operational planning that presents an alternative method of problem framing to facilitate better planning. Operational design is broken into seven sets of structured conversations that successively cover system framing, opponent rationale, command rationale, logistics rationale, operation framing, operational effects, and forms of function.
  - Government sponsors: Israeli Defense Forces, SOCOM, U.S. Army
  - Developer: Israeli Defense Forces' Operational Theory Research Institute
- Information
  - Focuses: Planning conceptualization and design
  - Operational level: Operational and strategic-level planning by Joint Chiefs of Staff, CCMDs, or joint task forces
  - Purpose: Rather than breaking a campaign or operational plan into smaller pieces to be analyzed individually, SOD embraces a systems thinking approach to operational and strategic-level planning by taking a holistic approach that focuses on the relationships between constituent parts. It is an iterative process that optimizes for a single objective but attempts to plan around a broad set of potentially competing objectives.
  - Forces: All
  - Current uses: Side-by-side comparison with traditional planning process during Unified Quest '05 (TRADOC), design of Senior Leader Seminars by SOCOM Wargame Center
  - Limitations: SOD is a departure from the traditional staff planning process and will require a certain level of buy-in, as well as increasing familiarity with the process. Additionally, a systems thinking process can quickly lead to the

---

[31] For more details, including background and templates, see William T. Sorrells et al., *Systemic Operational Design: An Introduction*, Fort Leavenworth, Kan.: School of Advanced Military Studies, U.S. Army Command and General Staff College, May 26, 2005.

consideration of factors outside military control, such as diplomacy or economics. A lack of interagency representation during the planning process can stymie this process.

- Requirements
  - Facilities: No special facilities. Large area for group discussion with available presentation and collaboration aids.
  - Equipment: Graphical or traditional text capture via Microsoft Word or similar programs to record discussion. Whiteboards and presentation aids for sketching relationships.
  - Personnel: Approximately six SOD-trained officers to form a design team
  - Required training: Officers who participated in SOD-style operational planning require two weeks of formal training and a month of independent study

Systems Thinking

- Overview[32]
  - Center: SOCOM
  - Tool/approach: Systems thinking
  - Description: An analysis methodology that emphasizes understanding the numerous factors affecting a situation, such as cycles of influence and one's own ability to intervene in a given situation
  - Government sponsor: N/A
  - Developer: Jay Forrester, Massachusetts Institute of Technology
- Information
  - Focuses: Visualization and design
  - Operational level: Any
  - Purpose: Visualization and wargame design
  - Forces: Any
  - Current uses: Used to dissect a complex problem into a manageable visual, which depicts how actors interact with each other, system strengths and weakness, and the relationship between various factors
  - Limitations: Causality can be oversimplified and visualization can easily get out of hand
- Requirements
  - Facilities: N/A
  - Equipment: N/A
  - Personnel: N/A
  - Required training: Minimal

---

[32] See Improvising Design, undated. For a guide to applying systems thinking toward wargaming, see Peter Perla and E. D. McGrady, *Systems Thinking and Wargaming*, Alexandria, Va.: Center for Naval Analyses, November 2009, a report that focuses on how to apply systems thinking in constructing various models of human organizations and processes, offering examples of various ways systems thinking can be applied to wargaming.

Tableau

- Overview[33]
  - Center: SOCOM Wargame Center
  - Tool/approach: Tableau
  - Description: A software tool that allows the visualization of large and complex data sets and multidimensional relational databases
  - Government sponsor: N/A; COTS product
  - Developer: Tableau Software
- Information
  - Focuses: Visualization and presentation
  - Operational levels: All
  - Purpose: To help senior-level decisionmakers gain better insight and understanding of the strategic impact of data
  - Forces: SOF
  - Current uses: Tableau has been used for visualizing the allocation of resources, maintenance schedules, and geospatial information
  - Limitations: Expense; potential permissions and licensing issues for multiple networks
- Requirements
  - Facilities: Space for large visual displays
  - Equipment: Tableau is available in desktop, server, and online forms. Requires Windows 7 or Windows Server 2008 or later.
  - Personnel: IT support for server installations
  - Required training: Free training videos are provided. Classes for end users and IT support staff run from two to three days.
  - Cost: $1,000–$2,000 for desktop version, $10,000 or more for server installations plus yearly support contracts

Tablets

- Overview
  - Center: CNA
  - Tool/approach: Tablet computers (iPad or other brands)
  - Description: Touch screen–enabled tablet computer useful for media display
  - Government sponsor: None
  - Developer: Apple (and others)
- Information
  - Focus: Presentation

---

[33] See Tableau, homepage, undated.

- Operational levels: All
- Purpose: iPads and other tablets can be used in place of paper printouts for scenario presentation to players and to provide injects that drive the game narrative. They can also be used by players as collaboration tools or by analysts to collect data.
- Forces: Any
- Current uses: Scenario presentation
- Limitations: None
- Requirements
  - Facilities: None
  - Equipment: iPad hardware
  - Personnel: None
  - Required training: Minimal

Technical Decision Support Wargame

- Overview
  - Center: DST
  - Tool/approach: Technical Decision Support Wargame
  - Description: Card-based game to identify future capabilities by structuring selection of packages of technologies to form capabilities and explore the capabilities used under different vignettes
  - Government sponsor: Dstl
  - Developer: Niteworks
- Information
  - Focuses: All
  - Operational level: Tactical
  - Purpose: Game design, COP, and adjudication
  - Forces: All
  - Current uses: Dstl
  - Limitations: Unclear
- Requirements
  - Facilities: Conference space
  - Equipment: Cards
  - Personnel: Facilitator and note takers
  - Required training: Unclear

Notes:
- Method has been used in one project to look at wide area persistent messaging, but appears to be readily adaptable.

ThinkTank
- Overview[34]
  - Center: J-8 SAGD
  - Tool/approach: ThinkTank
  - Description: Collective intelligence/group decision support software for brainstorming, innovation, decisionmaking, and virtual interactive meetings
  - Government sponsor: N/A
  - Developer: ThinkTank
- Information
  - Focus: Collaboration
  - Operational level: Strategic
  - Purpose: Facilitates brainstorming, consensus building, and stakeholder engagement. Helps game participants have better, richer discussions. Groups can contribute anonymously, and then collectively, to vet all possibilities. Teams can then evaluate and prioritize the results of their input. Finally, a meeting report is generated that contains clear deliverables and action items.
  - Forces: All
  - Current uses: J-8 SAGD has used ThinkTank software for its ability to pull the inputs from brainstorming sessions into quick surveys that participants can fill out and discuss
  - Limitations: May present a relatively steep learning curve for participants
- Requirements
  - Facilities: Conference room for participants, projector to display input data, interactive voting hardware
  - Equipment: Desktop computer with web browser
  - Personnel: N/A
  - Required training: N/A

Unified Engagement Technical Support
- Overview
  - Center: AFRL
  - Tool/approach: Unified Engagement technical support
  - Description: AFRL provides support representing emerging technology and capabilities into the Title 10 game Unified Engagement
  - Government sponsor: Air Force, as part of lead-up to Title 10 game series
  - Developer: AFRL
- Information
  - Focuses: Technology and capability assessment
  - Operational level: Operational

---

[34] See ThinkTank, homepage, undated.

- Purpose: Adjudication and postgame analysis
- Forces: All, but air focus
- Current uses: Air Force Title 10 inputs
- Limitations: Models and simulations do not have a good track record of performing during the game. Often they are too slow to use within the limited time available for adjudication tasks. Tools intended for adjudication often prove more helpful in postgame analysis, where they can provide an alternative perspective on adjudication choices made by SMEs during the game.
- Requirements
  - Facilities: Large gaming space to accommodate large number of teams. Air Force Title 10 games traditionally played at CCMDs, increasing logistical burden.
  - Equipment: Minimal
  - Personnel: Major line of effort for the center, which is five people
  - Required training: Unclear

Notes:
- Small macros implemented in Excel are often the most viable M&S tool for games, as they provide a tailored, easily revised tool for in-game adjudication.

Versatile Assessment Simulation Tool
- Overview
  - Center: TRAC
  - Tool/approach: VAST
  - Description: Visualization and adjudication support tool for wargaming. Able to "rewind" and readjudicate branches from the main wargame. Usable on a computer with a mouse and keyboard or on a touch screen device. Database attribute interaction–based tool.
  - Government sponsor: TRAC
  - Developer: TRAC
- Information
  - Focus: Conventional
  - Operational levels: Tactical and operational
  - Purposes: Adjudication, analysis, data collection, and visualization
  - Forces: Ground
  - Current uses: Wargame support
  - Limitations: Currently working on user's manual and unit input interface; can enter notes but not game-related documents
- Requirements
  - Facilities: No special facilities beyond what is required for the classification level of the work

- Equipment: PC or touch screen device
- Personnel: Four to five people to support a typical event; two working on software development, and two (variable) to support the event (depending on event size)
- Required training: One hour to learn to use the tool, but more time required to learn development

Notes:
- Would be sharable with the Marine Corps through an add-on to the existing memorandum of agreement that covers Combat XXI
- Uses correlation of forces and means to adjudicate combat results.

Videos/Video Studio
- Overview
  - Centers: CASL, CNA, NWC
  - Tool/approach: Videos/video studio
  - Description: Video filming and editing capabilities, ranging from a studio for live video production to camcorders and computers with editing software. Several centers stress the utility of being able to produce "new" video to deliver game scenario briefings and injects for heightened impact. Generally, in cases with more substantial capabilities, a center pulled on a shared video production capability, as the center by itself did not require full-time staff.
  - Government sponsors: CASL, NWC
  - Developers: CASL, NWC
- Information
  - Focuses: All
  - Operational level: Any
  - Purpose: COP
  - Forces: All
  - Current uses: Introduction briefings of game scenarios, scenario updates
  - Limitations: Video production can be expensive and time consuming, and the advantage compared with traditional briefing formats is contested. Often seen as being more important to nonmilitary senior leaders (including congresspeople and interagency leaders).
- Requirements
  - Facilities: Minimal (for lighter versions) to multiple custom-built studios (including soundproofing)
  - Equipment: Depending on center, ranges from commercial camcorder for filming original content and MacBook Pro for pulling available footage and editing a final product, to a full studio for filming and editing live footage

  – Personnel: No center required enough video to support a full-time person, but the time requirements were substantial to support production
  – Required training: Commercially available training for Adobe Creative Suite, iMovie, or similar editing software, ranging to fully trained production professionals

Notes:
  • The CNA has experimented with producing videos with virtual avatars instead of live actors in order to quickly edit videos on computer instead of reshooting with actors.

Virtual Worlds
  • Overview
    – Center: NUWC
    – Tool/approach: Virtual Worlds
    – Description: GOTS VR tool based on open-source software OpenSimulator. Virtual Worlds is a Navy-accredited version of OpenSimulator.
    – Government sponsor: U.S. Navy
    – Developer: GOTS version is based on open-source software OpenSimulator, which runs SecondLife
  • Information
    – Focus: Any
    – Operational levels: All
    – Purposes: Analysis, data collection, and visualization
    – Forces: Naval, ground, and air (current uses)
    – Current uses: Navy training, Navy wargaming support, Army training
    – Limitations: Unclear
  • Requirements
    – Facilities: Minimal
    – Equipment: One option is to maintain own instance on a standard server, and clients; second option is to rent space on NUWC server, access through the Navy Marine Corps Internet, and have NUWC maintain all software updates
    – Personnel: Individuals trained in Virtual Worlds
    – Required training: Two days of training modules developed by the Navy

Notes:
  • Business case of renting space on NUWC server may offer a very low startup cost for testing out the capability, especially since facility requirements are minimal
  • NUWC notes that acclimating to a VR world is important, but that actual technical skills required are not high. NUWC recommended training a junior person to be the equivalent of the "PowerPoint Ranger."

- Virtual Worlds was able to serve in knowledge management and wargame data collection and as a visualization platform. Demonstration encompassed everything from virtual re-creation of whiteboard and brainstorming exercise, to virtual prototyping of future submarine concepts, to wargame support, to use of external M&S tools to support wargame analysis.
- Has ability to replay text of wargame discussion as the animation steps through the wargame; could function as an extremely powerful visual outbrief method to senior decisionmakers on wargame results for ship-to-shore connectors and the like
- Able to handle federated DoD models and to connect to other external models (such as white shipping)
- The NUWC demonstration was at the tactical and system level for acquisition, but uses at other levels are certainly possible.

Wargame Infrastructure and Simulation Environment: Formation Wargame
- Overview
  - Center: Dstl
  - Tool/approach: Wargame Infrastructure and Simulation Environment: Formation Wargame
  - Description: Computer-based human-in-the-loop wargame and constructive simulation for battle group to division tactical actions, including air and maritime support. Stochastic, event driven model that allows players or software to make decisions.[35]
  - Government sponsor: Dstl
  - Developer: Dstl
- Information
  - Focus: Land forces
  - Operational levels: Battle group (battalion) through division
  - Purpose: Testing the effects of schemes of maneuver, capability, and force structure
  - Forces: Ground
  - Current uses: Dstl
- Requirements
  - Facilities: N/A
  - Equipment: PC
  - Personnel: Information unavailable
  - Required training: Information unavailable

---

[35] Paul Pearce, "The Wargaming Infrastructure and Simulation Environment," Defence Science and Technology Laboratory, Hampshire, UK, undated, pp. 1–2. Unpublished chapter provided by Dstl via email, February 16, 2018.

Notes:
- Fully closed wargame, with commanders only knowing what their forces have reported to them. Output is qualitative (e.g., insights supported by data) and quantitative (e.g., analysis of engagements).
- Primarily used to test the impact of changes on schemes of maneuver, capability, and force structures on the ability of battle group to division forces to complete the mission.

Zing Portable Team Meeting System
- Overview[36]
  - Center: DST
  - Tool/approach: Zing Portable Team Meeting System
  - Description: Software and hardware package designed for real-time collaboration, where users can anonymously project comments on a screen so they are not spoken over
  - Government sponsor: DST
  - Developer: Zing Technologies Pte Ltd
- Information
  - Focuses: Collaboration and knowledge management
  - Operational levels: All
  - Purpose: Facilitates freer communication between participants. Members can contribute via keyboard and have their comments displayed anonymously on a large screen so that all participants can participate regardless of rank or role. Includes support for popular voting methods and other participation tools.
  - Forces: Any
  - Current uses: Used by DST Joint and Operations Analysis Division during wargames to allow members to equally contribute
  - Limitations: Potential permissions and licensing issues for multiple networks
- Requirements
  - Facilities: Room large enough for all participants, with appropriate projection capabilities
  - Equipment: Desktop computer to run software and projection capabilities to display conversations. Package includes set of wireless keyboards, USB hubs and receivers, and appropriate software.
  - Personnel: One individual to set up
  - Required training: Minimal; includes instructions. System can be set up in minutes.

---

[36] See Working Visions, "Zing," webpage, undated.

Tables of Wargaming Tasks

This appendix contains tables of the core tasks that are common to any type of wargame and that we discuss in Chapter 6. The tables also contain a more complete list of tasks for the different wargaming types. Tasks that are specific to a given type of wargame or that are particularly emphasized in some way are differentiated within the tables.

Core Wargaming Tasks

Table C.1
Core Wargaming Tasks

Task	Description
Understanding sponsor requirements	Understanding the sponsor's objectives for the wargame and analysis.
Managing information	Acquiring and coordinating research among the wargame team.
Understanding the problem	Framing the problem to inform game design.
Developing and managing the event process	Coordinating logistical concerns involved with the game, such as venue, participants, and classification.
Scenario development	Crafting a scenario designed to inform and draw out the necessary decisions for participants within a game.
Game design	Creating and developing a system of decisionmaking and consequences within the game.
Game development	Developing a game through design and scenario validation, rule development, playtesting, and refinement.
Providing facilitation	Facilitating the game to keep players engaged and game events or discussion progressing.
Data capture	Capturing data and insights from the wargame, either through note takers or other tools.
Wargame analysis	Analyzing the wargame's insights, results, and date.

Concept Development Wargaming Tasks

Table C.2
Concept Development Wargaming Tasks

Wargaming Task	Description
Eliciting or identifying new concepts	Eliciting or identifying the new concepts to test within the wargame.
Understanding sponsor requirements	Understanding the sponsor's objectives for the wargame and analysis.
Managing information	Acquiring and coordinating research among the wargame team.
Understanding the problem	Framing the problem to inform game design.
Developing consistent assumptions about the future	Developing consistent assumptions about the future, such as the operating environment, technology, capabilities, and political climate.
Operationalizing concepts in the wargame	Creating opportunities in the game for players to use or respond to the identified concepts of interest. This may be done through representing new capabilities, units, different rules, or adjudication adjustments.
Developing and managing event process	Coordinating logistical concerns involved with the game, such as venue, participants, and classification.
Scenario development	Crafting a scenario designed to inform and draw out the necessary decisions for participants within a game.
Game development	Developing a game through design and scenario validation, rule development, playtesting, and refinement.
Game system design	Creating the processes, rules, mechanics, and visuals for a game.
Providing facilitation	Facilitating the game to keep players engaged and game events or discussion progressing.
Adjudicating novel concepts	Adjudicating new or novel concepts that are operationalized in the game but that may not exist in real life.
Identifying key uncertainties	Identifying the key uncertainties in a game and particularly uncertainties about the future and how a new or novel concept may be adjudicated.
Data capture	Capturing data and insights from the wargame, either through note takers or other tools.
Wargame analysis	Analyzing the wargame's insights, results, and data.

NOTE: Additional or emphasized tasks are in boldface and italics.

Capabilities Development and Analysis Wargaming Tasks

Table C.3
Capabilities Development and Analysis Wargaming Tasks

Wargaming Task	Description
Understanding sponsor requirements	Understanding the sponsor's objectives for the wargame and analysis.
Managing information	Acquiring and coordinating research among the wargame team.
Understanding the problem	Framing the problem to inform game design.
Retaining same fidelity as analytical model	Maintaining the wargame results at a high enough fidelity for data to be compatible with the appropriate analytical model or M&S tool.
Developing and managing event process	Coordinating logistical concerns involved with the game, such as venue, participants, and classification.
Scenario development	Crafting a scenario designed to inform and draw out the necessary decisions for participants within a game.
Game development	Developing a game through design and scenario validation, rule development, playtesting, and refinement.
Providing facilitation	Facilitating the game to keep players engaged and game events or discussion progressing.
Providing data-driven adjudication	Adjudicating the game based on data the DoD analysis community considers acceptable for programmatic decisionmaking.
Data capture	Capturing data and insights from the wargame, either through note takers or other tools.
Wargame analysis	Analyzing the wargame's insights, results, and data.
Generating data for further analysis	Creating data for further analysis to continue the cycle of research.

NOTE: Additional or emphasized tasks are in boldface and italics.

Science and Technology Wargaming Tasks

Table C.4
Science and Technology Wargaming Tasks

Wargaming Task	Description
Eliciting or identifying S&T concepts	Identifying S&T concepts to test within the wargame. Wargames rarely ever generate new concepts, but are useful in experimentation of new concepts.
Understanding sponsor requirements	Understanding the sponsor's objectives for the wargame and analysis.
Managing information	Acquiring and coordinating research among the wargame team.
Understanding the problem	Framing the problem to inform game design.
Developing consistent assumptions about the future	Developing consistent assumptions about the future, such as the operating environment, technology, capabilities, and political climate.
Developing consistent assumptions about S&T system or concept	When testing a novel S&R system or concept, designers must create a set of assumptions about its capabilities, capacity, and limitations.
Operationalizing concepts in the wargame	Creating opportunities in the game for players to use or respond to the identified concepts of interest. This may be done through representing new capabilities, units, different rules, or adjudication adjustments.
Developing and managing event process	Coordinating logistical concerns involved with the game, such as venue, participants, and classification.
Scenario development	Crafting a scenario designed to inform and draw out the necessary decisions for participants within a game.
Game development	Developing a game through design and scenario validation, rule development, playtesting, and refinement.
Game system design	Creating the processes, rules, mechanics, and visuals for a game.
Providing facilitation	Facilitating the game to keep players engaged and game events or discussion progressing.
Adjudicating novel concepts	Adjudicating new or novel concepts that are operationalized in the game but that may not exist in real life.
Identifying key uncertainties	Identifying the key uncertainties in a game and particularly uncertainties about the future and how a new or novel concept may be adjudicated.
Data capture	Capturing data and insights from the wargame, either through note takers or other tools.
Wargame analysis	Analyzing the wargame's insights, results, and data.

NOTE: Additional or emphasized tasks are in boldface and italics.

Senior Leader Engagement and Strategic Discussion Wargaming Tasks

Table C.5
Senior Leader Engagement and Strategic Discussion Wargaming Tasks

Wargaming Task	Description
Understanding sponsor requirements	Understanding the sponsor's objectives for the wargame and analysis.
Identifying key stakeholders	Involving key stakeholders, whether commanders, officials, or SMEs, in the wargaming process.
Managing information	Acquiring and coordinating research among the wargame team.
Understanding the problem	Framing the problem to inform game design.
Developing and managing event process	Coordinating logistical concerns involved with the game, such as venue, participants, and classification.
Accessing protocol officer	Accessing a protocol officer for logistical guidance when working with high-level officers or officials.
Scenario development	Crafting a scenario designed to inform and draw out the necessary decisions for participants within a game.
Game development	Developing a game through design and scenario validation, rule development, playtesting, and refinement.
Providing facilitation	Facilitating the game to keep players engaged and game events or discussion progressing.
Facilitating constructive discussion	Keeping senior leaders, in particular, on task and participating in constructive discussion.
Data capture	Capturing data and insights from the wargame, either through note takers or other tools.
Wargame analysis	Analyzing the wargame's insights, results, and data.

NOTE: Additional or emphasized tasks are in boldface and italics.

Operational Decisions and Plans Wargaming Tasks

Table C.6
Operational Decisions and Plans Wargaming Tasks

Wargaming Task	Description
Understanding sponsor requirements	Understanding the sponsor's objectives for the wargame and analysis.
Managing information	Acquiring and coordinating research among the wargame team.
Understanding the problem	Framing the problem to inform game design.
Defining operational environment	Scoping the operational environment in terms of context, capabilities, the adversary, and other operational factors.
Developing and managing event process	Coordinating logistical concerns involved with the game, such as venue, participants, and classification.
Scenario development	Crafting a scenario designed to inform and draw out the necessary decisions for participants within a game.
Game development	Developing a game through design and scenario validation, rule development, playtesting, and refinement.
Creating consistent and transparent adjudication process	Creating an adjudication process that is transparent and acceptable to operational planners.
Providing facilitation	Facilitating the game to keep players engaged and game events or discussion progressing.
Assessing real-life OPLAN	Assessing the strengths and weaknesses of the real-life OPLAN within the wargame.
Developing understanding of friendly and adversarial COAs	Developing and providing a realistic depiction of both friendly and adversarial COAs in terms of doctrine, capabilities, and capacity.
Data capture	Capturing data and insights from the wargame, either through note takers or other tools.
Wargame analysis	Analyzing the wargame's insights, results, and data.

NOTE: Additional or emphasized tasks are in boldface and italics.

Training and Education Wargaming Tasks

Table C.7
Training and Education Wargaming Tasks

Wargaming Task	Description
Understanding sponsor requirements	Understanding the sponsor's objectives for the wargame and analysis.
Understanding learning objectives and curriculum	Understanding the learning objectives and curriculum of the class and aligning the game with them.
Managing information	Acquiring and coordinating research among the wargame team.
Understanding the problem	Framing the problem to inform game design.
Drafting or co-opting game for learning objectives	Creating or adapting a game around learning objectives.
Developing and managing event process	Coordinating logistical concerns involved with the game, such as venue, participants, and classification.
Scenario development	Crafting a scenario designed to inform and draw out the necessary decisions for participants within a game.
Game development	Developing a game through design and scenario validation, rule development, playtesting, and refinement.
Providing facilitation	Facilitating the game to keep players engaged and game events or discussion progressing.
Providing immersive experience	Providing an immersive experience in which students must assume the roles they play within a game. Gameplay should be engaging and fun.
Data capture	Capturing data and insights from the wargame, either through note takers or other tools.
Providing feedback to students	Providing feedback to students on their gameplay.
Evaluating learning	Evaluating the level of student learning from a game in relation to the learning objectives and the class curriculum.
Wargame analysis	Analyzing the wargame's insights, results, and data.

NOTE: Additional or emphasized tasks are in boldface and italics.

Bibliography

Abt, Clark C., *Serious Games*, New York: Viking Press, 1970.

AFMC—*See* Air Force Materiel Command.

Air Force Materiel Command, "AFMC History" and "AFMC Mission," webpage, undated. As of April 26, 2017:
http://www.afmc.af.mil/Home/Welcome/

Air University, "Curtis E. LeMay Center for Doctrine Development and Education," webpage, undated. As of April 27, 2017:
http://www.au.af.mil/au/lemay/main.htm

———, "U.S. Air Force Wargaming Gateway," webpage, March 12, 2019. As of July 6, 2019:
http://www.airuniversity.af.mil/lemay/display/article/1099721/us-air-force-wargaming-gateway-mil-only/

Allen, Thomas B., *War Games: The Secret World of the Creators, Players, and Policy Makers Rehearsing World War III Today*, New York: McGraw-Hill, 1987.

Altman, Howard, "SOCOM Seeking Latest Tech, Even Holograms, for Wargame Center," *Tampa Bay Times*, March 26, 2017.

Appleget, Jeff, "Analytic Wargaming: Introduction to Wargaming and Best and Worst Practices," briefing, Military Operations Research Society, Monterey, Calif., June 2015.

———, "Module 1" and "Module 2," annotated teaching slides, Naval Postgraduate School, Monterey, Calif., August 2016.

Appleget, Jeff, and Robert Burks, "Naval Postgraduate School Mobile Training Team (MTT) Wargaming Program," briefing slides, Naval Postgraduate School, Monterey, Calif., August 2016.

Appleget, Jeff, and Fred Cameron, "Analytic Wargaming on the Rise," *Phalanx*, Vol. 48, No. 1, March 2015, pp. 28–32.

Appleget, Jeff, Fred Cameron, and Robert E. Burks, "Wargaming at the Naval Postgraduate School," special issue, "Modeling and Simulation Special Edition: Wargaming," *CSIAC Journal*, Vol. 4, No. 3, November 2016, pp. 18–23. As of July 6, 2019:
https://www.csiac.org/wp-content/uploads/2016/12/CSIAC_Journal_V4N3_Nov2016.pdf

ArcGIS, homepage, undated. As of July 6, 2019:
https://www.arcgis.com/index.html

ATLAS.ti, homepage, undated. As of June 17, 2019:
https://atlasti.com

Australian Department of Defence, Defence, Science and Technology, "About DST," webpage, undated. As of July 6, 2019:
https://www.dst.defence.gov.au/discover-dsto/about-dst

Bartels, Elizabeth. "Innovative Education: Gaming—Learning at Play," *ORMS Today*, Vol. 41, No. 4, August 2014. As of July 6, 2019:
https://www.informs.org/ORMS-Today/Public-Articles/August-Volume-41-Number-4/INNOVATIVE-EDUCATION-Gaming-Learning-at-play

Bestard, Jaime J., "Air Force Research Laboratory Innovation: Pushing the Envelope in Analytical Wargaming," special issue, "Modeling and Simulation Special Edition: Wargaming," *CSIAC Journal*, Vol. 4, No. 3, November 2016, pp. 12–17. As of July 6, 2019:
https://www.csiac.org/wp-content/uploads/2016/12/CSIAC_Journal_V4N3_Nov2016.pdf

Birkner, Christine. "From Monopoly to Exploding Kittens, Board Games are Making a Comeback." *Adweek*, April 3, 2017. As of July 18, 2019:
https://www.adweek.com/brand-marketing/from-monopoly-to-exploding-kittens-board-games-are-making-a-comeback/

Body, Howard, and Colin Maston, "The Peace Suppose Operations Model: Origins, Development, Philosophy and Use," *Journal of Defense Modeling and Simulation*, Vol. 8, 2010, pp. 69–77.

Brewer, Garry D., and Martin Shubik, *The War Game: A Critique of Military Problem Solving*, Cambridge, Mass.: Harvard University Press, 1979.

Brown, Tim, and Jocelyn Wyatt, "Design Thinking for Social Innovation IDEO," *Stanford Social Innovation Review*, Winter 2010. As of July 6, 2019:
https://ssir.org/articles/entry/design_thinking_for_social_innovation

Brynen, Rex, "Playtesting RCAT," PAXsims (blog), November 30, 2015. As of October 27, 2018:
https://paxsims.wordpress.com/2015/11/30/playtesting-rcat/

Burns, Shawn, ed., *War Gamers' Handbook: A Guide for Professional War Gamers*, Newport, R.I.: Naval War College, Wargaming Department, 2013.

CAA—*See* Center for Army Analysis.

Caffrey, Matthew B., Jr., "Completing the Wargame Cycle: Event & Item Wargames," briefing, Headquarters, Air Force Materiel Command, Wright-Patterson Air Force Base, Ohio, January 11, 2017a.

———, "Wargaming: Design & Development," briefing, Headquarters, Air Force Materiel Command, Wright-Patterson Air Force Base, Ohio, January 11, 2017b.

———, *On Wargaming: How Wargames Have Shaped History and How They May Shape the Future*, Naval War College, Newport Papers 43, Washington, D.C.: U.S. Government Publishing Office, 2019.

Cameron, F., and G. Pond, *Military Decision Making Using Schools of Thought Analysis—A Soft Operational Research Technique, with Numbers*, Ottawa, Ont., Canada: Defence Research and Development Canada, Centre for Operational Research and Analysis, 2010.

Campbell, Colin, "How Fortnite's Success Led to Months of Intense Crunch at Epic Games," *Polygon*, April 23, 2019. As of April 24, 2019:
https://www.polygon.com/2019/4/23/18507750/fortnite-work-crunch-epic-games

Carter, Clarence E., Phillip D. Coker, and Stanley Gorene, *Dynamic Commitment: Wargaming Projected Forces Against the QDR Defense Strategy*, Washington, D.C.: Institute for National Strategic Studies, Strategic Forum No. 131, November 1997.

CASL—*See* Center for Applied Strategic Learning.

Center for Applied Strategic Learning, "CASL History," webpage, undated. As of May 30, 2017:
http://casl.ndu.edu/About/About-CASL/

Center for Army Analysis, homepage, undated. As of May 30, 2017:
http://www.caa.army.mil

Center for Naval Analyses, "CNA's Center for Naval Analyses," webpage, undated. As of May 9, 2017:
https://www.cna.org/centers/cna/

Centre for Operational Research and Analysis, electronic comments, September 20, 2017.

Cisco, "Cisco Jabber," webpage, undated. As of June 17, 2019:
http://www.cisco.com/web/products/voice/jabber.html

CNA—*See* Center for Naval Analyses.

Connections UK, "Aim and Purpose," webpage, undated. As of September 4, 2017:
http://www.professionalwargaming.co.uk/About.html

Connections Wargaming Conference, homepage, undated. As of September 4, 2017:
https://connections-wargaming.com/

Costikyan, Greg, "The Unfulfilled Promise of Digital Wargames," in Pat Harrigan and Matthew G. Kirschenbaum, eds., *Zones of Control: Perspectives on Wargaming*, Cambridge, Mass.: MIT Press, 2016, pp. 681–689.

Curedale, Robert, *Design Thinking: Process and Methods*, 2nd ed., Topanga, Calif.: Design Community College Inc., 2016.

Curry, John, and Tim Price, *Matrix Games for Modern Wargaming Developments in Professional and Educational Wargames: Innovations in Wargaming*, Vol. 2, Morrisville, N.C.: Lulu Press, 2014.

Decision Lens, homepage, undated. As of June 17, 2019:
https://www.decisionlens.com

Defence Science and Technology Laboratory, *Dstl's Wargaming Team*, brochure, June 27, 2016.

DOD—*See* U.S. Department of Defense.

Dstl—*See* Defence Science and Technology Laboratory.

Ducharme, Douglas, "Approaches to Title 10 Wargaming," Newport, R.I.: Naval War College, Wargaming Department, undated.

Dunnigan, James F., *Wargames Handbook: How to Play and Design Commercial and Professional Wargames*, 3rd ed., Bloomington, Ind.: iUniverse, 2000.

Email correspondence with Naval Undersea Warfare Center staff, July 18, 2017.

Email from Defence Science and Technology Laboratory staff, February 16, 2018.

Email from Office of Naval Intelligence staff, September 19, 2017.

Email from Office of the Secretary of Defense, Cost Assessment and Performance Evaluation staff, September 19, 2017.

Email from Special Operations Command Wargame Center staff, October 5, 2017.

Email from U.S. Army War College staff, September 19, 2017.

Email interview with Military Operations Research Society wargaming community of practice member, June 29, 2017a.

Email interview with Military Operations Research Society wargaming community of practice member, July 6, 2017b.

Email interview with U.S. Navy wargaming community of practice member, July 17, 2017.

Email interview with U.S. Navy wargaming community of practice member, April 24, 2019.

Email correspondence with Naval War College staff, August 2, 2017.

Facilitate.com, homepage, undated. As of July 14, 2019:
https://www.facilitate.com

Gabel, Christopher R., *The U.S. Army GHQ Maneuvers of 1941*, Washington, D.C.: U.S. Army Center of Military History, 1991.

The Game Crafter, "MaGCK Matrix Game Construction Kit," webpage, undated. As of June 17, 2019:
https://www.thegamecrafter.com/games/magck-matrix-game-construction-kit

Goldhamer, Herbert, and Hans Speier, *Some Observations on Political Gaming*, Santa Monica, Calif.: RAND Corporation, P-1679-RC, 1959. As of June 17, 2019:
https://www.rand.org/pubs/papers/P1679.html

Gorak, Mark, "Introduction," special issue, "Modeling and Simulation Special Edition: Wargaming," *CSIAC Journal*, Vol. 4, No. 3, November 20, 2016, pp. 5–7. As of July 7, 2019:
https://www.csiac.org/wp-content/uploads/2016/12/CSIAC_Journal_V4N3_Nov2016.pdf

Gould, Mark, *The CAEn Process—Wargaming, Simulation and Replication*, London: UK Ministry of Defence, Defence Science and Technology Laboratory, September 8, 2016. As of October 27, 2018:
http://professionalwargaming.co.uk/TheCAEnProcess.pdf

Hagel, Chuck, U.S. Secretary of Defense, "The Defense Innovation Initiative," memorandum, Washington, D.C., November 15, 2014.

Hambling, David, "Game Controllers Driving Drones, Nukes," *Wired*, July 19, 2008. As of July 7, 2019:
https://www.wired.com/2008/07/wargames/

Hanley, John Thomas, Jr., *On Wargaming: A Critique of Strategic Operational Gaming*, dissertation, Yale University, 1991.

Harvard Extension School, "Creative Thinking: Innovative Solutions to Complex Challenges," webpage, undated. As of July 7, 2019:
http://dev.extension.harvard.edu/professional-development/programs/creative-thinking-innovative-solutions-complex-challenges

Hay, Bud, and Bob Gile, *Global War Game*, Newport, R.I.: Naval War College Press, 1993.

Headquarters, U.S. Army Training and Doctrine Command, *Organizations and Functions: United States Army TRADOC Analysis Center*, Fort Monroe, Va.: TRADOC Regulation 10-5-7, July 2010.

Headquarters, U.S. Department of the Army, *The Operations Process*, Washington, D.C.: Army Doctrine Reference Publication ADRP 5-0, May 2012.

Headquarters, U.S. Marine Corps, *Marine Corps Capabilities Based Assessment*, Washington, D.C.: Marine Corps Order 3900.20, September 27, 2016.

———, *Marine Corps Planning Process*, Washington, D.C.: Marine Corps Warfighting Publication MCWP 5-10, April 4, 2018.

Heuer, Richard J., Jr., and Randolph H. Pherson, *Structured Analytic Techniques for Intelligence Analysis*, 2nd ed., Thousand Oaks, Calif.: CQ Press, 2015.

Hofman, Rudalf M., *War Games*, n.p., Historical Division, Headquarters, U.S. Army Europe, 1952.

Hopkin, Tony, "Dstl Wargaming Support to the Army," briefing, Defence Science and Technology Laboratory, Salisbury, England, September 14, 2014.

IBM, "IBM i2 Analyst's Notebook," webpage, undated. As of June 17, 2019:
https://www.ibm.com/us-en/marketplace/analysts-notebook

Improvising Design, homepage, undated. As of October 4, 2017:
http://www.improvisingdesign.com/

Intelligence Advanced Research Projects Activity, "Using Alternate Reality Environments to Help Enrich Research Efforts (UAREHERE)," IARPA-RFI-13-03, Washington, D.C., March 11, 2013. As of July 18, 2019:
https://www.iarpa.gov/index.php/working-with-iarpa/requests-for-information/using-alternate
-reality-environments-to-help-enrich-research-efforts

International Association of Facilitators, homepage, undated. As of June 17, 2019:
https://www.iaf-world.org/site/

Interview with Center for Applied Strategic Learning staff, Fort McNair, Washington, D.C., August 26, 2016.

Interview with Center for Army Analysis wargamers, Fort Belvoir, Va., December 12, 2016.

Interview with Center for Naval Analyses staff, Arlington, Va., January 4, 2016.

Interview with Command and General Staff College staff, Fort Leavenworth, Kan., August 20, 2016.

Interview with J8 Studies, Analysis, and Gaming Division staff, Arlington, Va., November 21, 2016.

Interview with members of the Marine Corps Operations Analysis Directorate, Quantico, Va., March 23, 2017.

Interview with members of the Marine Corps Wargaming Division, Quantico, Va., March 23, 2017.

Interview with Naval Postgraduate School faculty, Monterey, Calif., August 15, 2016.

Interview with Naval Undersea Warfare Center staff, Newport, R.I., August 4, 2016.

Interview with Naval War College Wargaming Department staff, Newport, R.I., August 3, 2016.

Interview with Office of the Secretary of Defense Cost Assessment and Performance Evaluation staff, Arlington, Va., October 17, 2016.

Interview with University of Foreign Military and Cultural Studies staff, Fort Leavenworth, Kan., August 30, 2017.

Interview with video game industry analyst, Los Angeles, Calif., September 4, 2016.

John Tiller Software, homepage, undated. As of July 14, 2019:
http://www.johntillersoftware.com

Joint Chiefs of Staff, *Joint Capabilities Integration and Development System (JCIDS)*, Washington, D.C.: Chairman of the Joint Chiefs of Staff Instruction 3170.01I, January 23, 2015, p. A-3.

———, *Joint Planning*, Washington, D.C.: Joint Publication JP 5-0, June 16, 2017a.

———, *Joint Planning*, Washington, D.C.: Joint Publication JP 5-0, 2017b.

Joint Staff, J-8, "J-8 Force Structure, Resource & Assessment," webpage, undated. As of May 30, 2017:
https://www.jcs.mil/Directorates/J8-Force-Structure-Resources-Assessment/

Jolin, Dan, "The Rise and Rise of Tabletop Gaming," *Guardian*, September 25, 2016.

Kendall, Frank, Under Secretary of Defense for Acquisition, Technology and Logistics, "Long Range Research and Development Plan (LRRDP) Direction and Tasking," memorandum, Washington, D.C., October 29, 2014.

King's College London News Centre, "New Wargaming Network Launched at King's," press release, London, December 3, 2018. As of April 23, 2019:
https://www.kcl.ac.uk/news/new-wargaming-network-launched

Koopman, Bernard Osgood, *Search and Screening: General Principles with Historical Applications*, Alexandria, Va.: Military Operations Research Society, 1999.

Lacey, James, "Wargaming in the Classroom: An Odyssey," *WarontheRocks.com*, April 19, 2016. As of July 9, 2019:
https://warontherocks.com/2016/04/wargaming-in-the-classroom-an-odyssey/

Leadership Strategies, homepage, undated. As of June 17, 2019:
https://www.leadstrat.com

Levine, Robert A., Thomas Schelling, and William M. Jones, *Crisis Games 27 Years Later: Plus C'est Déjà vu*, Santa Monica, Calif.: RAND Corporation, P-7719, 1991. As of July 10, 2019:
https://www.rand.org/pubs/papers/P7719.html

Mahoney, Daniel P., "Center for Army Analysis Wargame Analysis Model (C-WAM)," PowerPoint presentation for the Military Operations Research Society Wargaming Community of Practice, Arlington, Va., April 20, 2016a.

———, *The Joint Wargaming Analysis Model*, version 8.1, Fort Belvoir, Va.: Center for Army Analysis, November 2016b.

Maurer, John, "USMC Wargaming Capability," PowerPoint presentation, Marine Corps Systems Command, Quantico, Va., January 24, 2019.

Maybus, Ray, U.S. Secretary of the Navy, "Wargaming," memorandum, Washington, D.C., May 5, 2015.

Mayer, Igor, "The Gaming of Policy and the Politics of Gaming: A Review," *Simulation & Gaming*, Vol. 40, No. 6, 2009, pp. 825–862.

MCCDC/CD&I—*See* U.S. Marine Corps Combat Development Command / Combat Development and Integration.

McCown, Margaret, "Defense Wargaming Panel: The SAGD/Provider Perspective," presentation at the Connections Wargaming Conference, Quantico, Va., August 2017.

McCue, Brian, *Wotan's Workshop: Military Experiments Before the Second World War*, Washington, D.C.: Center for Naval Analyses, 2002.

McHugh, Francis J., *Fundamentals of War Gaming*, 3rd ed., Newport, R.I.: U.S. Naval War College, Strategic Research Department, 1966.

Meas, Corey, *War Play: Video Games and the Future of Armed Conflict*, New York: Houghton Mifflin Harcourt, 2013.

Menadue, Ion, Scott James, and Lance Holden, "jSWAT2: The Application of Simulation to Support Seminar Wargaming," PowerPoint presentation, Defence Science and Technology, n.p., Australia, undated a.

———, "jSWAT2: The Application of Simulation to Support Seminar Wargaming," Defence Science and Technology, n.p., Australia, undated b. As of July 24, 2019:
http://www.simulationaustralasia.com/files/upload/pdf/research/56-09.pdf

Mid-Atlantic Facilitator Network, homepage, undated. As of June 17, 2019:
https://www.mafn.org

Military Operations Research Society, "About MORS," webpage, undated a. As of August 24, 2017:
http://www.mors.org/Home/About

———, "Communities of Practice," webpage, undated b. As of August 24, 2017:
http://www.mors.org/Communities

———, *MORS Special Meeting: Cyberspace Special Meeting—Wargaming & Analytics, Terms of Reference*,
Arlington, Va.: Military Operations Research Society, October 17, 2018. As of April 23, 2019:
https://www.mors.org/Portals/23/Docs/Events/2018/Cyberspace%20Wargaming/181003%20
%202018%20%20MORS%20CSMDMW%20TOR.pdf

"Modeling and Simulation Special Edition: Wargaming," special issue, *CSIAC Journal*, Vol. 4, No. 3,
November 2016. As of July 6, 2019:
https://www.csiac.org/wp-content/uploads/2016/12/CSIAC_Journal_V4N3_Nov2016.pdf

MORS—*See* Military Operations Research Society.

Morse, Philip M., and George E. Kimball, *Methods of Operations Research*, Mineola, N.Y.: Dover
Publications, 2003.

National Defense University, *Annual Report for Academic Year 2016 (AY16)*, Washington, D.C.:
National Defense University, 2016.

Naval Postgraduate School, "NPS Vision," webpage, undated. As of June 21, 2017:
https://my.nps.edu/vision

Naval Sea Systems Command, "Warfare Centers: NUWC Newport Division," webpage, undated. As
of June 22, 2017:
http://www.navsea.navy.mil/Home/Warfare-Centers/NUWC-Newport/Who-We-Are/

NDU—*See* National Defense University.

Newzoo Games, *Free 2016 Global Games Market Report: An Overview of Trends & Insights*, San
Francisco: Newzoo Games, June 2016. As of September 4, 2017:
https://cdn2.hubspot.net/hubfs/700740/Reports/Newzoo_Free_2016_Global_Games_Market_
Report.pdf

Nofi, Albert A., *To Train the Fleet for War, 1923–1940*, Newport, R.I.: Naval War College Press,
2010.

NPS—*See* Naval Postgraduate School.

NUWC—*See* Naval Undersea Warfare Center.

NWC— *See* U.S. Naval War College.

Office of Naval Intelligence, "Our Mission," webpage, undated. As of May 30, 2017:
http://www.oni.navy.mil/Our_Mission/

Office of the Secretary of Defense, Cost Assessment and Program Evaluation, "About CAPE,"
webpage, undated. As of May 30, 2017:
http://www.cape.osd.mil

———, "Wargaming in CAPE," briefing at Connections 2017, Quantico, Va., June 1, 2017.

ONI—*See* Office of Naval Intelligence.

OpenSimulator, homepage, undated. As of July 13, 2019:
http://opensimulator.org/wiki/Main_Page

OSD CAPE—*See* Office of the Secretary of Defense, Cost Assessment and Program Evaluation.

Pardee RAND Graduate School, "Center for Gaming," webpage, undated. As of July 7, 2019:
https://www.prgs.edu/research/methods-centers/gaming.html

Parson, Edward A., "What Can You Learn From a Game?" in Richard Zeckhauser, Ralph L. Keeney, and James K. Sebenius, eds., *Wise Choices: Games, Decisions, and Negotiations*, Boston, Mass.: Harvard Business School Press, 1996.

Pearce, Paul, "The Wargaming Infrastructure and Simulation Environment," Defence Science and Technology Laboratory, Hampshire, UK, undated. Unpublished chapter provided by Dstl via email, February 16, 2018.

Perla, Peter, *Peter Perla's The Art of Wargaming: A Guide for Professionals and Hobbyists*, ed. John Curry, Morrisville, N.C.: Lulu Press, 2012.

Perla, Peter, and E. D. McGrady, *Systems Thinking and Wargaming*, Alexandria, Va.: Center for Naval Analyses, November 2009.

———, "Why Wargaming Works," *Naval War College Review*, Vol. 64, No. 1, Summer 2011. As of July 7, 2019:
https://digital-commons.usnwc.edu/cgi/viewcontent.cgi?article=1578&amp=&context=nwc-review

Pincombe, Brandon, Sarah Blunden, Adrian Pincombe, and Patricia Dexter, "Ascertaining a Hierarchy of Dimensions from Time-Poor Experts: Linking Tactical Vignettes to Strategic Scenarios," *Technological Forecasting and Social Change*, Vol. 80, No. 4, 2013, pp. 584–598.

Plattner, Hasso, *Introduction to Design Thinking Process Guide*, Stanford, Calif.: Institute of Design at Stanford, 2010. As of July 7, 2019:
https://dschool-old.stanford.edu/sandbox/groups/designresources/wiki/36873/attachments/74b3d/ModeGuideBOOTCAMP2010L.pdf

Polski, Margaret, "Are Wargames Quasi-Experiments?" panel discussion at the 83rd MORS Annual Symposium, Alexandria, Va., June 24, 2015.

Pournelle, Phillip E., "Designing Wargames for the Analytic Purpose," *Phalanx*, Vol. 50, No. 2, June 2017a, pp. 48–53.

———, "MORS Wargaming Workshop III," *Phalanx*, Vol. 50, No. 2, June 2017b, pp. 27–29.

———, ed., *MORS Wargaming Special Meeting, October 2016: Final Report*, Alexandria, Va.: Military Operations Research Society, 2017c.

Pournelle, Phillip, and Holly Deaton, eds., *MORS Wargaming III Special Meeting, 17–19 October 2017: Final Report*, Alexandria, Va.: Military Operations Research Society, April 2018. As of April 23, 2019:
https://www.mors.org/Portals/23/Docs/Events/2017/Wargaming/MORS%20Wargaming%20III%20Report%20Final.pdf

R Foundation, "The R Project for Statistical Computing," webpage, undated. As of July 7, 2019:
https://www.r-project.org

RAND Corporation, "Wargaming," webpage, undated. As of May 30, 2017:
https://www.rand.org/topics/wargaming.html

The Research and Analysis Center, "TRAC as a Federal Laboratory," webpage, undated. As of May 30, 2017:
http://www.trac.army.mil/about.html

———, "Synthetic Staff Ride (SSR): Mindanao," presentation for the Military Operations Research Society, Alexandria, Va., May 13, 2015.

———, "The Athena Simulation," briefing, Fort Leavenworth, Kan., August 12, 2016a.

———, "Wargaming Introduction and Applications," presentation for the RAND Corporation, Santa Monica, Calif., August 31, 2016b.

Robinson, Merle, Stephen Downes-Martin, and Connections US Wargaming Conference 2018 Working Group, *In-Stride Adjudication*, Washington, D.C.: National Defense University, July 19, 2018. As of April 24, 2019:
https://paxsims.files.wordpress.com/2018/09/in-stride-adjudication-working-group-report-20180908.pdf

Ross, M. Brian, U.S. Naval War College, "Naval Wargaming Becomes a Collaborative Community," press release, Washington, D.C., June 2017. As of August 24, 2017:
http://www.secnav.navy.mil/innovation/Documents/2017/06/WargamingCommunity.pdf

Sabin, Philip, *Simulating War: Studying Conflict Through Simulation Games*, New York: Continuum International Publishing Group, 2012.

Shubik, Martin, *Games for Society, Business, and War: Towards a Theory of Gaming*, New York: Elsevier, 1975.

Sinclair, Brendan. "Digital Market Jumps 16% in July—Superdata," *Gamesindustry.biz*, August 24, 2017. As of August 30, 2017:
http://www.gamesindustry.biz/articles/2017-08-24-digital-market-jumps-16-percent-in-july-superdata

SOCOM—*See* U.S. Special Operations Command.

Sorrells, William T., Glen R. Downing, Paul J. Blakesley, David W. Pendall, Jason K. Walk, and Richard D. Wallwork, *Systemic Operational Design: An Introduction*, Fort Leavenworth, Kan.: School of Advanced Military Studies, U.S. Army Command and General Staff College, May 26, 2005.

South, Todd, "Marine Wargaming Center Will Help Plan for Future Combat," *Marine Corps Times*, September 19, 2017. As of April 23, 2019:
https://www.marinecorpstimes.com/news/your-marine-corps/2017/09/19/marine-wargaming-center-will-help-plan-for-future-combat/

Stanford University, "Hacking for Defense: Class Details," webpage, undated a. As of June 17, 2019:
http://hacking4defense.stanford.edu/details.html

———, d.school, "Tools for Taking Action," webpage, undated b. As of June 17, 2019:
https://dschool.stanford.edu/resources/

Strong, Paul Edward, "Manual Wargaming Approaches at Dstl," briefing, Defence Science and Technology Laboratory, Salisbury, England, September 18, 2014.

Tableau, homepage, undated. As of June 17, 2019:
https://www.tableau.com

Telephone conversation with LeMay Center for Doctrine Development and Education personnel, April 4, 2017.

Telephone conversation with U.S. Army War College staff, November 18, 2016.

Telephone interview with Centre for Operational Research and Analysis staff, August 23, 2016.

Telephone interview with Defence Science and Technology Laboratory staff, September 8, 2017.

Telephone interview with Defence Science and Technology staff, November 23, 2016.

Telephone interview with Office of Naval Intelligence staff, October 14, 2016.

Telephone interview with Special Operations Command Wargame Center staff, November 2, 2016.

Telephone interview with wargamers at Air Force Materiel Command and the Air Force Research Laboratory, October 4, 2016.

ThinkTank, homepage, undated. As of June 17, 2019:
https://thinktank.net

Thomas, Joe, Air Force Global Strike Command, "GE 16 Keeps AFGSC on the Edge of Deterrence," press release, Barksdale Air Force Base, La., June 22, 2016. As of May 30, 2017:
http://www.afgsc.af.mil/News/Article-Display/Article/809954/ge-16-keeps-afgsc-on-the-edge-of-deterrence/

Thomas, William, *Rational Action: The Science of Policy in Britain and America, 1940–1960*, Cambridge Mass.: MIT Press, 2015.

TRAC—*See* The Research and Analysis Center.

TRADOC—*See* U.S. Army Training and Doctrine Command.

UFMCS—*See* University of Foreign Military and Cultural Studies.

UK Ministry of Defence, Development, Concepts and Doctrine Centre, *Defence Wargaming Handbook*, Swindon, England: UK Ministry of Defense, August 2017.

UK MOD—*See* UK Ministry of Defence.

University of Foreign Military and Cultural Studies, *Liberating Structures Handbook*, Fort Leavenworth, Kan.: University of Foreign Military and Cultural Studies, undated a.

———, *UFMCS Group Think Mitigation Guide*, Fort Leavenworth, Kan.: University of Foreign Military and Cultural Studies, undated b.

———, *The Applied Critical Thinking Handbook*, version 8.1, Fort Leavenworth, Kan.: University of Foreign Military and Cultural Studies, September 2016.

U.S. Army Combined Arms Center, "Mission, Vision, Priorities, Principles, & College-Level Learning Outcomes: About the Command and General Staff College," webpage, undated a. As of July 20, 2017:
http://usacac.army.mil/organizations/cace/cgsc/mission

———, "University of Foreign Military and Cultural Studies / Red Teaming," webpage, undated b. As of August 21, 2017:
http://usacac.army.mil/organizations/ufmcs-red-teaming

U.S. Army Modeling and Simulation Office, "Military Program—FA57," webpage, undated. As of July 20, 2017:
http://www.ms.army.mil/sp-div2/fa57/index.html

U.S. Army Training and Doctrine Command, "About TRADOC," webpage, undated. As of July 7, 2019:
https://www.tradoc.army.mil/About-Us/

U.S. Army War College, "About the US Army War College," webpage, undated a. As of August 19, 2017:
http://www.armywarcollege.edu/overview.htm

———, "The Center for Strategic Leadership," webpage, undated b. As of August 20, 2017:
http://csl.armywarcollege.edu

———, "Department of Strategic Wargaming (DSW)," webpage, undated c. As of August 19, 2017:
http://csl.armywarcollege.edu/DSW.aspx

———, "January–June, 2017: International Fellows (IFs) Matrix-Game," webpage, undated d. As of July 13, 2019:
https://csl.armywarcollege.edu/archives.aspx

USAWC—*See* U.S. Army War College.

U.S. Central Intelligence Agency, *A Tradecraft Primer: Structured Analytic Techniques for Improving Intelligence Analysis*, Washington, D.C.: U.S. Central Intelligence Agency, May 2009.

U.S. Department of Defense, *Support for Strategic Analysis*, Washington, D.C.: Department of Defense Directive DoDD 8260.05, July 7, 2011.

U.S. Department of Defense, Under Secretary of Defense (Comptroller), *Financial Management Regulation*, Washington, D.C.: Department of Defense Directive DoD 7000.14-R, June 2017.

U.S. Department of the Navy, Navy Warfare Development Command, *Navy Planning*, Washington, D.C.: Navy Warfare Publication NWP 5-01, December 2013.

U.S. Marine Corps, Wargaming Division, Connections Wargaming Conference 2017 briefing, Quantico, Va., August 2, 2017.

U.S. Marine Corps Combat Development Command / Combat Development and Integration, *Concept Development*, Quantico, Va.: Marine Corps Combat Development Command/Combat Development and Integration Instruction 5401.1, February 8, 2016.

———, *Initial Capability Document for Marine Corps Wargaming Capability*, version 1.3, Quantico, Va., May 2017. Not available to the general public.

U.S. Naval War College, "About U.S. Naval War College," webpage, undated a. As of July 18, 2016: https://www.usnwc.edu/About.aspx

———, "Our Mission," webpage, undated b. As of July 7, 2019: https://usnwc.edu/About/Mission

———, *War Gaming: United States Naval War College*, brochure, Newport, R.I.: U.S. Naval War College, undated c. As of July 7, 2019: https://web.archive.org/web/20170627235754/https://www.usnwc.edu/getattachment/e32b4fba-9daf-4462-9d32-d8a7875f2abb/War-Gaming-Brochure.aspx

U.S. Special Operations Command, "Headquarters USSOCOM," webpage, undated a. As of July 13, 2019: https://www.socom.mil/about

———, homepage, undated b. As of July 13, 2019: https://www.socom.mil/

Vaidos, Alexandra, review of ACH, Softpedia, undated. As of July 8, 2019: https://www.softpedia.com/get/Science-CAD/ACH.shtml

Vebber, Paul, "The Battle to Teach Wargaming," *Wargaming Connection Blog*, May 8, 2016. As of May 30, 2017: https://wargamingcommunity.wordpress.com/2016/05/08/the-battle-to-teach-wargaming

Vego, Milen, "German War Gaming," *Naval War College Review*, Vol. 65, No. 4, Autumn 2012. As of July 9, 2019: https://digital-commons.usnwc.edu/cgi/viewcontent.cgi?article=1494&context=nwc-review

WarfareSims.com, "Command: Modern Air/Naval Operations," webpage, undated a. As of June 17, 2019: http://www.warfaresims.com/?page_id=1101

———, homepage, undated b. As of July 14, 2019: http://www.warfaresims.com

Webb, Kevin, "How Nintendo's Handheld Video Game Consoles Have Evolved over the Past 30 Years, from the Original Game Boy to the Switch," *Business Insider*, April 23, 2019. As of April 24, 2019: https://www.businessinsider.com/nintendo-game-boy-history-evolution-ds-3ds-switch-2019-4

Weiner, Milton G., *Trends in Military Gaming*, Santa Monica, Calif.: RAND Corporation, P-4173, 1969. As of June 17, 2019:
https://www.rand.org/pubs/papers/P4173.html

Wenger, Etienne, *Communities of Practice: Learning, Meaning, and Identity*, New York: Cambridge University Press, 1998.

———, "Communities of Practice: A Brief Introduction," paper presented to the STEP Leadership Workshop, University of Oregon, October 2011.

WGD—*See* U.S. Marine Corps, Wargaming Division.

Williams, Peter B., and Fred D. J. Bowden, "Dynamic Morphological Exploration," paper presented at the 22nd National Conference of the Australian Society for Operations Research, Adelaide, Australia, December 1–6, 2013.

Wilson, Andrew, *The Bomb and the Computer*, New York: Delacorte Press, 1969.

Work, Bob, and Paul Selva, "Revitalizing Wargaming Is Necessary to Be Prepared for Future Wars," *WarontheRocks.com*, December 8, 2015:
https://warontherocks.com/2015/12/revitalizing-wargaming-is-necessary-to-be-prepared-for-future-wars/

Work, Robert, and U.S. Deputy Secretary of Defense, "Wargaming and Innovation," memorandum, Washington, D.C., February 9, 2015.

Working Visions, "Zing," webpage, undated. As of June 17, 2019:
http://www.workingvisions.com.au/Zing.htm

Wright-Patterson Air Force Base, "Air Force Research Laboratory," webpage, undated. As of May 29, 2017:
http://www.wpafb.af.mil/afrl/

Zenko, Micah, "Millennium Challenge: The Real Story of a Corrupted Military Exercise and Its Legacy," *WarontheRocks.com*, November 5, 2015. As of July 10, 2019:
https://warontherocks.com/2015/11/millennium-challenge-the-real-story-of-a-corrupted-military-exercise-and-its-legacy/